Top Drugs:
Top Synthetic Routes

John Saunders

Vice President of Chemistry at Neurocrine Biosciences,
San Diego, California

Series sponsor: **ZENECA**

ZENECA is a major international company active in four main areas of business:
Pharmaceuticals, Agrochemicals and Seeds, Speciality Chemicals, and Biological Products.

ZENECA's skill and innovative ideas in organic chemistry and bioscience create products
and services which improve the world's health, nutrition, environment, and quality of life.

ZENECA is committed to the support of education in chemistry and chemical engineering.

OXFORD
UNIVERSITY PRESS

OXFORD
UNIVERSITY PRESS

Great Clarendon Street, Oxford OX2 6DP

Oxford University Press is a department of the University of Oxford.
It furthers the University's objective of excellence in research, scholarship,
and education by publishing worldwide in

Oxford   New York

Athens   Auckland   Bangkok   Bogotá   Buenos Aires   Calcutta
Cape Town   Chennai   Dar es Salaam   Delhi   Florence   Hong Kong   Istanbul
Karachi   Kuala Lumpur   Madrid   Melbourne   Mexico City   Mumbai
Nairobi   Paris   Saõ Paulo   Singapore   Taipei   Tokyo   Toronto   Warsaw

with associated companies in   Berlin   Ibadan

Oxford is a registered trade mark of Oxford University Press
in the UK and in certain other countries

Published in the United States
by Oxford University Press Inc., New York

A catalogue record for this book is available from the British Library

Library of Congress Cataloging in Publication Data
(Data applied for)
ISBN 0 19 850100 5
Typeset by EXPO Holdings, Malaysia
Printed in Great Britain
on acid-free paper by Bath Press Ltd., Bath, Avon

Series Editor's Foreword

Mankind has benefitted immeasurably from the contribution that organic synthesis has made to the discovery and development of pharmaceuticals. The discovery of new lead compounds and the optimisation of their structure to elicit the best pharmacological profile, thus generating novel drug candidates, requires the invention of versatile synthetic routes and methodologies. In this Primer John Saunders describes synthetic routes that have been used to synthesise the structures of 37 of the top drugs in current usage. This provides an ideal format for introducing students to a wide range of applied chemistry alongside brief descriptions of the modes of action of these drugs.

Oxford Chemistry Primers have been designed to provide concise introductions relevant to all students of chemistry and contain only the essential material that would normally be covered in an 8-10 lecture course. In this Primer John Saunders provides an interesting and very readable account of this enormously important area. This Primer will be of interest to apprentice and Master chemist alike.

Stephen G. Davies
The Dyson Perrins Laboratory, University of Oxford.

Author's Preface

Today's modern drugs have been uncovered from two major sources: natural products and laboratory synthesis. For the most part, medicines derived from the former source have allowed the diversity of nature's biosynthetic machinery to produce either the compound directly or an advanced intermediate which subsequently may be manipulated in the laboratory to give the desired spectrum of activity. Those synthesised directly by medicinal chemists usually have been the result of a protracted discovery programme using a natural product (e.g. a hormone or an enzyme substrate) or a screening lead as a starting point. Where a drug has been produced by post-biosynthetic modification, it has been excluded from the current manuscript because, frequently, the biosynthesis and synthesis(es) have been reviewed in detail elsewhere.

The original intention of this book was to compare the initial research route to a new compound with that of the final manufacturing process. The latter information drug companies clearly want to keep to themselves. Despite multiple inquiries to most major companies whose products are included in this book, there was little willingness to help. Nevertheless, it has been educational to study the original syntheses with more recent approaches which aim either at improving the route or at validating newer methodologies or reagents in the context of drug synthesis. Since for many drugs the marketed product was originally prepared as a racemic mixture, perhaps the most important comparison is between that route and alternatives which involve some element of asymmetric synthesis.

The chapters are ordered into many of the major categories of human disease—cardiovascular (Chapters 1–3), gastrointestinal (Chapters 4–5), central nervous system (Chapters 6–7), inflammatory disease (Chapter 8) and infectious diseases (Chapters 9–10). Individual chapters are focussed on collections of drugs that operate via the same mechanism. After an introduction to the discovery and mechanism of action, each drug is discussed in terms of its first synthesis and then later routes which have points of interest to the organic chemist. It is important to note, that, unlike many 'academic' syntheses, the best methods are not necessarily the most elegant, rather the most pragmatic. One interesting challenge to the reader is to determine which route may be the most convenient on a manufacturing scale and what changes should be made to address any deficiencies in this respect. To help, consider the availability and cost of starting materials and reagents, safety of the proposed reactions using reaction vessels of hundreds of gallons, reproducibility of the process, disposal of the waste materials, the possibility of multi-step, one pot reactions and sensitivity of the yield to minor changes in reaction conditions. These issues are discussed in detail in 'Process Development' by S. Lee and G. Robinson, Oxford Chemistry Primer No. 30. Finally, some key references are given to the literature describing each drug, usually the US patent citation and one or more pivotal paper; note that the reference listing is meant not to be comprehensive.

2000   J.S.

Contents

1	Inhibitors of angiotensin converting enzyme as effective antihypertensive agents	1
2	Blockade of angiotensin-II receptors	16
3	Calcium channel blockers in the treament of angina and hypertension	21
4	Antagonists of histamine receptors (H_2) as anti-ulcer remedies	31
5	Proton pump inhibitors as gastric acid secretion inhibitors	37
6	Modulation of central serotonin in the treatment of depression	43
7	Hypnotic, anxiolytic, anticonvulsant and muscle relaxant agents: ligands for benzodiazepine receptors	52
8	Another histamine receptor: blockers of the H_1 receptor for the treament of seasonal allergic rhinitis	61
9	Nucleoside analogues which inhibit HIV reverse transcriptase	71
10	Quinolones as anti-bacterial DNA gyrase inhibitors	80
	Abbreviations	88
	Index	89

1 Inhibitors of angiotensin converting enzyme as effective antihypertensive agents

1.1 Introduction

During the 1970's, the mainstays for the treatment of hypertension were diuretics, beta-blockers or a combination of the two. Whilst often effective, these treatments were not without serious side effects such as electrolyte imbalance and lethargy respectively. For some time prior to this period, there had been considerable interest in the role of the renin-angiotensin (RA) system (Scheme 1.1) in the control of blood pressure but there had been no real advances for two reasons: (1) It had not been possible to design non-peptide inhibitors of any step of the cascade although peptide analogues of angiotensin-II had been demonstrated to lower blood pressure in hypertensive animals. (2) Many groups had the notion that essential hypertension was mediated through the RA system in only a small proportion of patients and

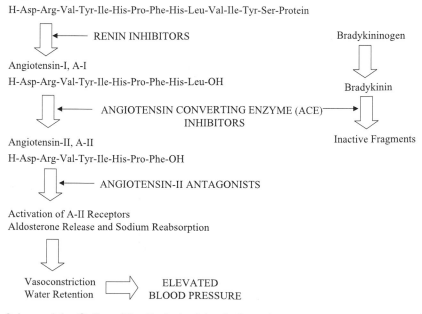

Angiotensinogen
H-Asp-Arg-Val-Tyr-Ile-His-Pro-Phe-His-Leu-Val-Ile-Tyr-Ser-Protein

RENIN INHIBITORS

Bradykininogen

Angiotensin-I, A-I
H-Asp-Arg-Val-Tyr-Ile-His-Pro-Phe-His-Leu-OH

Bradykinin

ANGIOTENSIN CONVERTING ENZYME (ACE) INHIBITORS

Inactive Fragments

Angiotensin-II, A-II
H-Asp-Arg-Val-Tyr-Ile-His-Pro-Phe-OH

ANGIOTENSIN-II ANTAGONISTS

Activation of A-II Receptors
Aldosterone Release and Sodium Reabsorption

Vasoconstriction
Water Retention

ELEVATED
BLOOD PRESSURE

Scheme 1.1.   Outline of the Renin-Angiotensin Cascade.

had lost interest. The discovery of captopril **1**, a landmark in rational drug design, changed all that.

Angiotensin converting enzyme (ACE) plays a central role in a cascade of proteolytic reactions which ultimately control the levels of angiotensin II, a potent vasoconstrictor (A-II Scheme 1.1). Starting with angiotensinogen, renin, an aspartyl proteinase synthesised by the kidney but secreted into the plasma, specifically cleaves this protein substrate at the Leu[10]-Val[11] amide bond to produce A-I in the rate limiting step. The resulting decapeptide is further cleaved at Phe[8]-His[9] by ACE acting as a carboxydipeptidase to generate A-II which can elevate blood pressure either directly by activating A-II receptors located on smooth muscle cells in arterioles or indirectly by stimulating the release of aldosterone from the kidney. In turn, aldosterone causes sodium reabsorption in the kidneys and water retention thereby increasing blood volume and pressure. Intervention at any point on this cascade should reduce blood pressure in the hypertensive patient. Despite heroic efforts spread over a decade of research, orally active inhibitors of renin proved elusive; in contrast, ACE inhibitors provided an exciting breakthrough

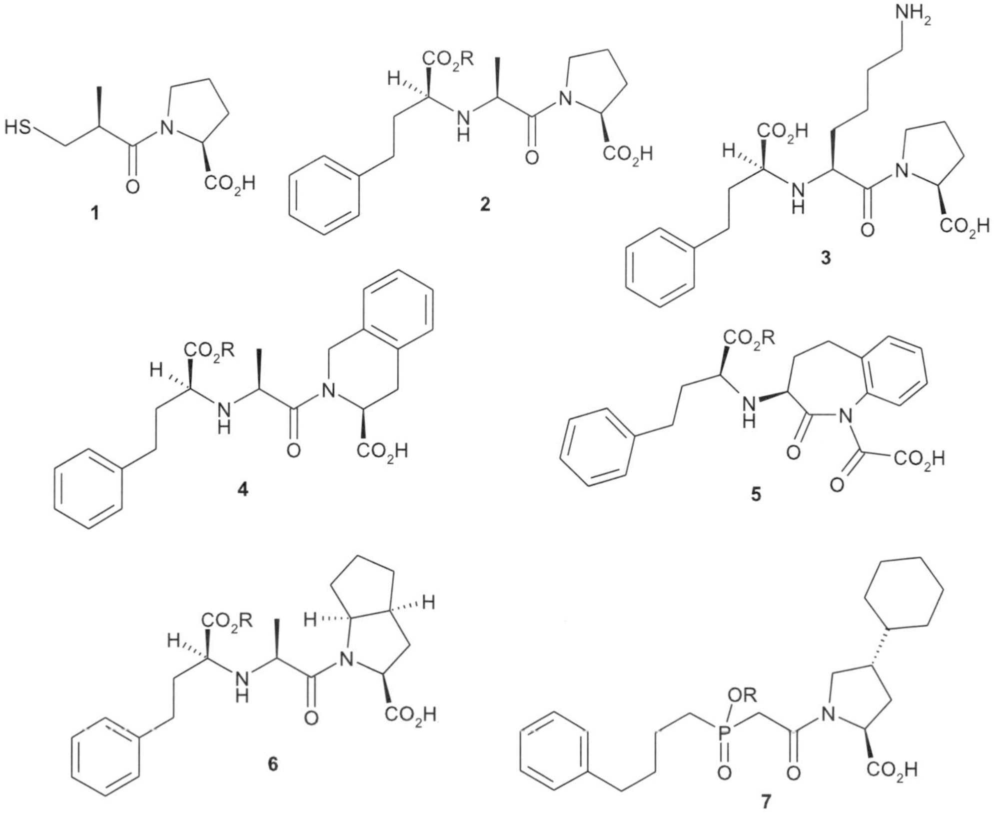

Scheme 1.2. Structures of leading ACE inhibitors. For **2, 4, 5 and 6**, R = H represents the active drug whereas R = Et is the prodrug form designed to improve oral bioavailability. For **7**, the active drug has R = H but the prodrug form is R = CH{CH(CH$_3$)$_2$}OCOEt.

in rational drug design In passing, it should be noted that ACE also plays a role in the regulation of levels of the vasodilatory agent, bradykinin; side effects such as cough and angioedema have been attributed to elevated levels of this hormone. Thus, it has been suggested that antagonists of A-II at the receptor level may well have the same efficacy in lowering blood pressure but devoid of these unwanted side effects (see Chapter 2).

Early clinical trials with captopril were difficult to administer – as often happens with the first candidate of a completely novel mechanistic class. Much of the negative points of view focused on the presence of the sulphydryl group in captopril and competitors raced to find an alternative group which could ligand to the active site Zn. The resulting carboxylic acid based inhibitors, enalapril **2** (R = Et; acting as a prodrug form of the active acid, R = H), lisonopril **3**, quinapril **4**, benazepril **5** and ramipril **6**, and the phosphinic acid, fosinopril **7** (R = CH(CH(CH$_3$)$_2$)OCOEt, active form has R = H) have also become highly successful products.

1.2   Synthesis of captopril

In a pivotal paper demonstrating rational drug design, the active site of ACE was likened to that of carboxypeptidase A (CPA) making due allowance for the fact that the former is a carboxydipeptidase (the enzyme cleaves the peptide substrate two residues from the C-terminus). Noting an observation that D-2-benzylsuccinic acid was a potent inhibitor of CPA, it was postulated to represent a biproduct inhibitor (Scheme 1.3) with the acetic acid entity interacting with the active site Zn^{2+} atom. Extending the inhibitor by one

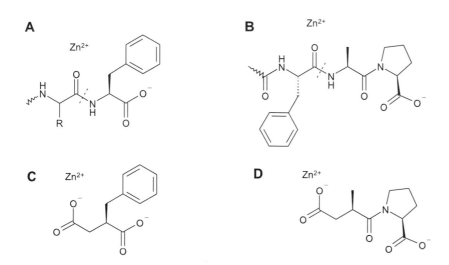

Scheme 1.3.   Design of ACE inhibitors (see text). A: C-terminus of CPA substrate with scissile amide bond carbonyl aligned with active site Zn; B: C-terminus of A-I, the substrate for ACE; C: benzylsuccinic acid, an inhibitor of CPA; D: prototypic inhibitor of ACE.

amino acid residue and taking into account the substrate specificity differences between the two enzymes, the prototypic ACE inhibitor was born. Building on this logic by substituting the carboxylic acid with a thiol group having far greater avidity for Zn^{2+}, led to the discovery of captopril.

The synthesis of captopril is inherently simple. 3-Acetylthio-2-methylpropionic acid **8** was coupled with L-proline t-butyl ester (Scheme 1.4) in the presence of dicyclohexylcarbodiimide (DCCI). The crude product was taken into TFA and anisole and the resulting acid **9** purified as the dicyclohexylamine salt. The material that was isolated by crystallisation from acetonitrile afforded the (S,S) diastereomer whilst the (R,S) form could be obtained from the mother liquors. The thiol group was unmasked by treatment of the separated thioacetates with methanolic ammonia being careful to exclude the presence of oxygen to avoid disulphide formation. Notably, the (S,S)-isomer **1** with a $K_i = 1.7$ nM was about 100 fold more active than the (R,S) isomer.

Scheme 1.4. Reagents: (a) hydroquinone, 90°, 1 h; (b) DCCI, DCM, rt, 16 h; (c) TFA, PhOMe, rt, 1 h; (d) separate diastereomers as DCHA salt; (e) MeOH, NH₃, rt, 2 h.

Alternative syntheses have focussed on the need to avoid the separation of diastereomers **9** or resolution of **8**. One approach used the microbiological induced stereoselective hydration of methacrylic acid to give **10** (Scheme 1.5) as the only product. This material was then elaborated into captopril without compromising the configurational integrity of the asymmetric center and without the need for protection of the thiol group.

Scheme 1.5. Reagents: (a) microbiological hydration; (b) SOCl₂, imidazole, 10°, then 80°, 3 h; (c) L-proline, 2N NaOH, H₂O, 1 h; (d) NaSH, DMF, 4 h, 50° under N₂.

The asymmetric synthesis of captopril was also used as a vehicle to demonstrate the use of the chiral iron auxilliary **11** (Scheme 1.6) which is commercially available in the chiral pseudo octahedral form. Thus, the enantiomerically pure (R-configuration) acetyl complex may be readily deprotonated with n-BuLi at −78° and the enolate quenched with methyl iodide to give **12**. The orientation of the acyl group places the O atom antiperiplanar to the carbon monoxide ligand with the alkyl chain lying parallel to a phenyl ring of the Ph$_3$P moiety. A second deprotonation reaction afforded the E-enolate **13** upon treatment with n-BuLi at low temperature and this was captured with bromomethyl-t-butyl sulphide acting exclusively at the unhindered face to give only the (R,S)-complex **14**. Addition of bromine and then L-proline t-butyl ester gave **15** which was deprotected with TFA to yield isomerically pure (S,S)-captopril.

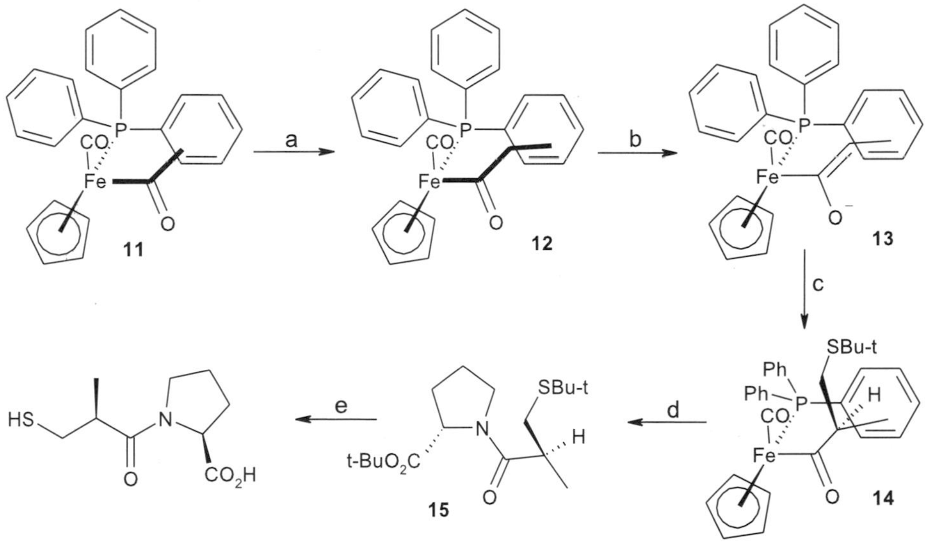

Scheme 1.6. Reagents: (a) n-BuLi, THF, −78°, then MeI; (b) n-BuLi, THF, −78°; (c) t-BuSCH$_2$Br; (d) Br$_2$ then proline t-butyl ester; (e) TFA, Hg(OAc)$_2$, then H$_2$S.

1.3 Synthesis of Enalaprilat (2, R = H) and Lisinopril (3)

Enalaprilat was first prepared as a diastereomeric mixture by reductive coupling of 2-oxo-4-phenylbutyric acid with the dipeptide Ala-Pro in the presence of sodium cyanoborohydride (Scheme 1.7). Inorganic bi-products were removed by absorption of the product on to Dowex-50 in the acid form. Separation of the isomers was achieved by chromatography on XAD-2 polystyrene resin eluting with 0.1 M NH$_4$OH in water and methanol. The (S,S,S)-enantiomer is almost 1000 times more active than the (R,S,S) epimer indicating that a highly defined lipophilic pocket is responsible accommodating the phenethyl sidechain.

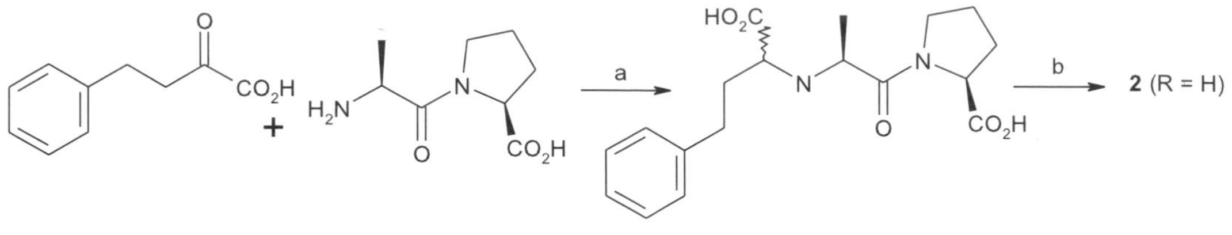

Scheme 1.7   Reagents: (a) NaCNBH$_3$, H$_2$O, rt, 16 h; then Dowex-50 X-2; (b) separate diastereomers on XAD-2 polystyrene resin.

Following literature observations describing the asymmetric synthesis of α-amino acids, a more efficient synthesis of enalapril was devised (Scheme 1.8). An excess of the α-keto ester was condensed with Ala-Pro to form the Schiff base **16** in the presence of a reducing reagent to avoid epimerisation of this intermediate. Catalytic reduction, but not reduction with NaCNBH$_3$, gave enalapril enriched in the SSS diastereomer (62:38). Formation and recrystallisation of the maleic acid salt gave the desired SSS isomer in greater than 99% purity in 32% overall yield.

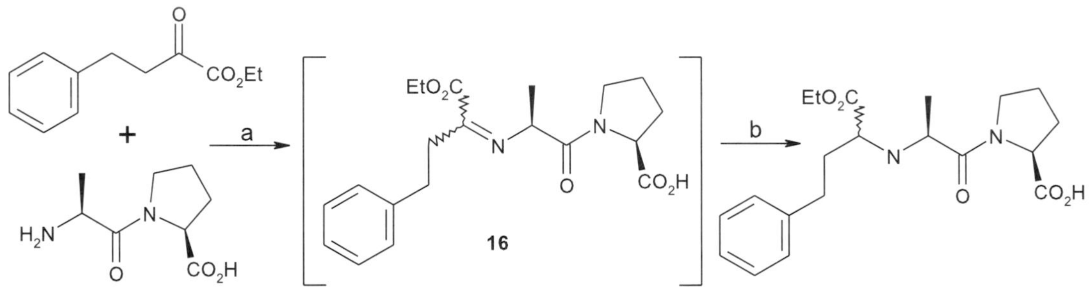

Scheme 1.8.   Reagents: (a) EtOH, 4A mol sieve; (b) then 10% Pd-C, 40 psi, rt, 15 h then MeCN, maleic acid, fractional crystallisation to separate diastereomers.

Because of the functionalised sidechain in lisinopril (4-aminobutyl replaces the 'alanine' methyl of enalapril), a protecting group strategy had to be used for this molecule (Scheme 1.9). After the amide coupling step, reductive alkylation again gave a diastereomeric mixture which, after deprotection with TFA, had to be separated chromatographically on a XAD-2 polystyrene column.

1.4   Quinapril

A somewhat different approach involving direct N-alkylation rather than reductive alkylation and a re-ordering of the coupling sequence characterises the synthesis of quinapril (Scheme 1.10) in which a tetrahydroquinoline ring replaces the proline ring of earlier ACE inhibitors. Thus the pseudo-dipeptide **17** was available as a mixture of diastereomers by alkylation of S-alanine t-butyl ester followed by deprotection and these isomers were readily separated

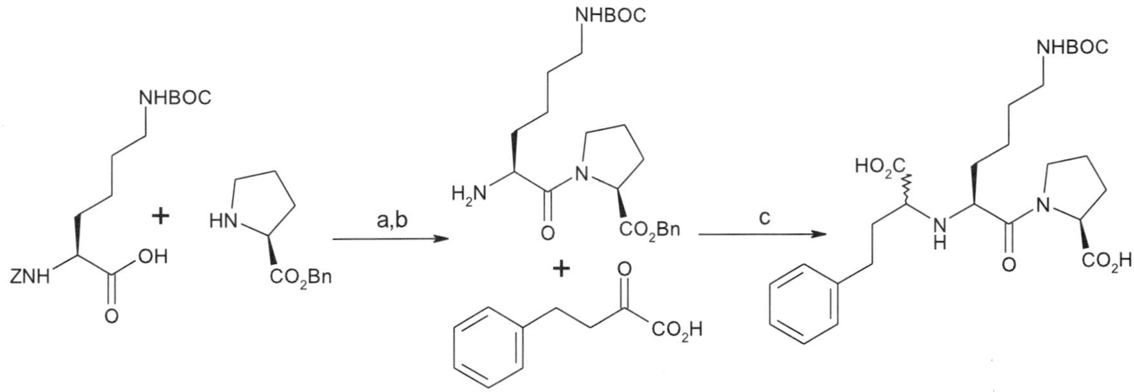

Scheme 1.9. Reagents: (a) DCC, Et$_3$N, DCM, rt, 16 h; (b) H$_2$, 10% Pd-C, EtOH; HOAc, rt, 15 h, 40 psi; (c) H$_2$O, NaOH to pH 7.0, NaCNBH$_3$, rt, 24 h.

by fractional crystallisation. The (S,S)-isomer was coupled with the S-form of tetrahydro-3-isoquinolinecarboxylic acid t-butyl ester **18**, prepared by a Pictet Spengler reaction on L-phenylalanine followed by esterification of the acid with isobutylene, to afford **19**. Both quinapril and quinaprilat were accessible by sequential deprotection first with acid and then with aqueous base. It was necessary to avoid prolonged storage of the intermediate monoethyl ester as the free base since under certain conditions the diketopiperazine **20** was found to contaminate the product.

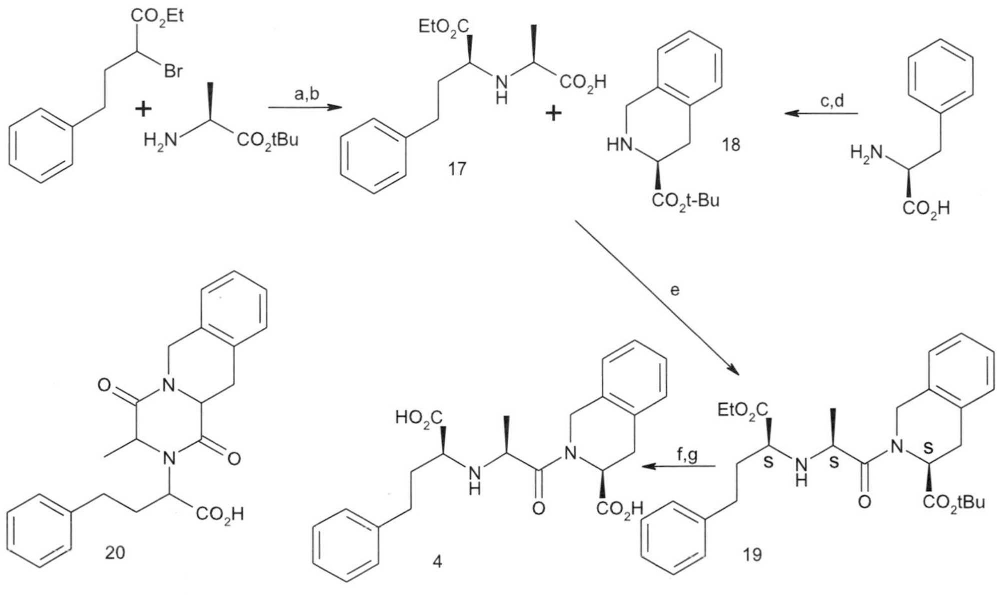

Scheme 1.10. Reagents: (a) DMF, Et$_3$N, 70°, 18 h; (b) TFA, rt, 1 h then separate diastereomers as HCl salts (the required (S,S)-isomer from mother liquors); (c) aq. CH$_2$=O, reflux, then HCl; (d) isobutylene, conc. H$_2$SO$_4$, dioxane, rt, 24 h; (e) HBT, DCCI, THF, Et$_3$N, 5°, 1 h; (f) TFA, rt, 1 h; (g) 1N NaOH, MeOH, rt, 20 h.

1.5 Benazepril

Benazepril **5** evolved from thiol based inhibitors **21** in the (mistaken) belief that the thiol group of captopril was responsible for the side effect profile seen in earlier clinical studies with that drug. As indicated above, it was subsequently shown that captopril is effective at much lower doses than those prescribed initially and any side effects seem more to be mechanism based independent of chemical structure.

The benzazepinone starting material **22** (Scheme 1.11) was readily prepared from α-tetralone by acid catalysed addition of hydrazoic acid in a Schmidt ring enlargement reaction. α-Dichlorination followed by controlled reduction gave the mono chlorolactam **23** and this afforded the azide **24** upon treatment with NaN$_3$. N-alkylation under phase transfer conditions and reduction of the azide gave **25** (R = Et) as a racemic mixture of enantiomers which was resolved as the tartrate salt. The (−)-isomer of **25** was tentatively assigned (S)-configuration by correlation with the ring contracted analogue obtained from L-tryptophan which also had (−) optical rotation. This isomer was first hydrolysed with sodium hydroxide and the resulting amino acid

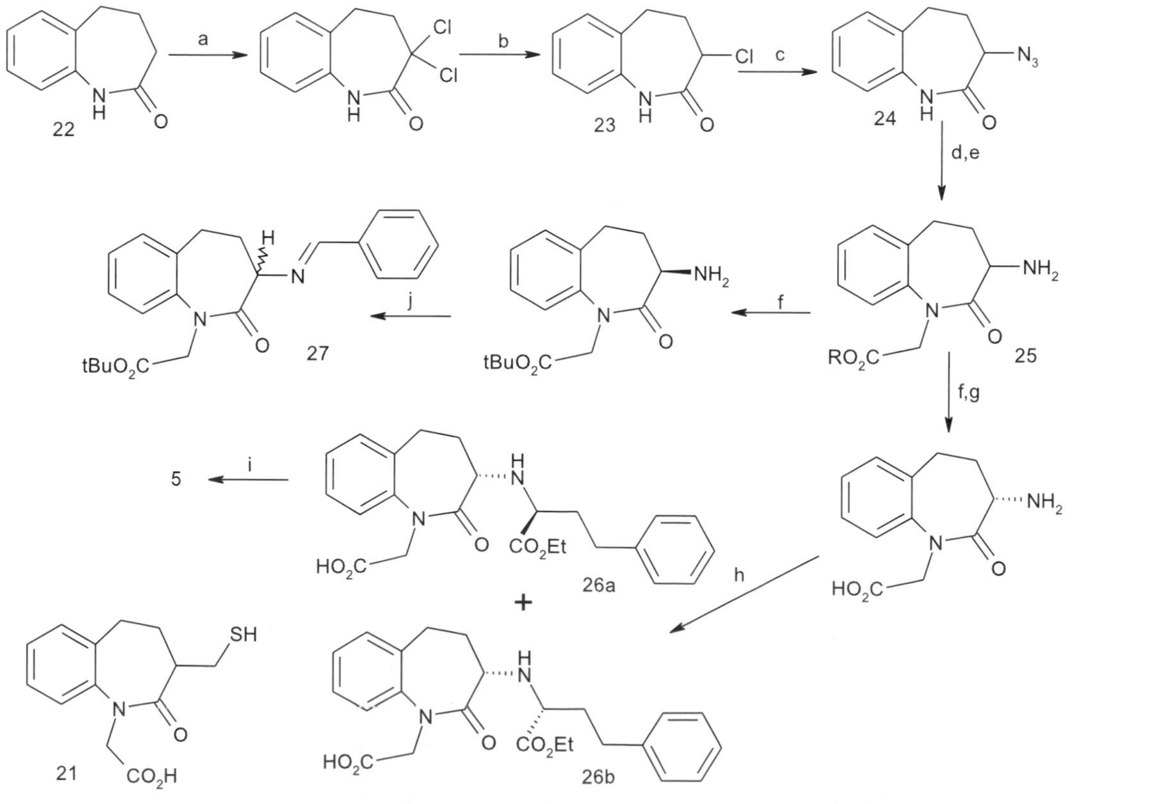

Scheme 1.11. Reagents: (a) PCl$_5$, 90°, 0.5 h; (b) HOAc, NaOAc, 5% Pd-C; (c) Me$_2$SO, NaN$_3$, 80°, 3 h; (d) Bu$_4$N$^+$Br$^-$, KOH, BrCH$_2$CO$_2$Et, or BrCH$_2$CO$_2$^{t}Bu, rt, 1.5 h; (e) EtOH, 10% Pd-C; (f) resolve as tartrate salt, EtOH; (g) R = Et, MeOH, NaOH, H$_2$O, rt, 2 h; (h) Ph(CH$_2$)$_2$COCO$_2$Et, HOAc, MeOH, rt, 1 h then NaCNBH$_3$, 16, rt, separate; (i) NaOH, H$_2$O, MeOH, 18 h rt; (j) PhCHO.

(sodium salt) reductively alkylated with ethyl benzylpyruvate. The major product **26a** from a 7:3 ratio of diastereomers was purified by crystallisation and finally hydrolysed to benazeprilat, the active (S,S)-form of **5**. Subsequently, an enantioselective synthesis involving **25** (R = t-Bu) emerged. After resolution of the intermediate amine **25** it proved possible to recycle the unwanted isomer by first forming the imine **27** with benzaldehyde to induce epimerisation of the stereogenic center followed by hydrolysis of the imine.

1.6 Ramipril

Like other ACE inhibitors with the aminocarboxylate pharmacophore, ramipril is rapidly converted by hepatic cleavage of the ethyl ester entity to ramiprilat, the active diacid. Having five stereogenic centers, the molecule is potentially the most complex within this class; the drug is sold as a single enantiomer, notably the (S,S,S,3aS,6aS) isomer wherein the bicyclic system has cis, endo-configuration although the exo-isomer (S,S,S,3aR,6aR) is also a potent inhibitor.

The octahydrocyclopenta[b]pyrrole core of ramipril was available by several methods, three of which will act to illustrate the chemistry. The first started with 2-oxo-cyclopentyl acetic acid **28** (Scheme 1.12) which was reductively cyclised with Raney nickel in the presence of ammonia to afford the lactam **29** having cis stereochemistry at the ring junction. This was reduced to the secondary amine and then oxidised to the imine with $Hg(OAc)_2$. Addition of cyanide gave **30** which, as the trifluoroacetate, could be separated into cis, endo and cis, exo racemates. Acid hydrolysis of the nitrile and esterification gave the ethyl ester **31** which was used as an alternative opportunity for isomer separation. Exclusively the endo form of the same key intermediate was accessible from the benzylimine of cyclopentanone in two

Scheme 1.12. Reagents: (a) Ra-Ni, NH_3; (b) $LiAlH_4$, THF; (c) $Hg(OAc)_2$, HOAc, H_2O, reflux, 6 days; (d) KCN, MeOH, H_2O, HCl, 16 h; (e) separate epimers as trifluoroacetate derivative; (f) 6N HCl, reflux, 8 h; (g) HCl, EtOH, $(EtO)_3CH$, reflux 16 h; (h) $BrCH_2CO_2Et$, EtOH, Et_3N; (i) 20% $Pd(OH)_2$, EtOH.

Scheme 1.13. Reagents: (a) PCl$_5$; (b) DMF, 24 h, rt; (c) 2N HCl, reflux, 1 h; (d) H$_2$, 10% Pd-C, HOAc; (e) SOCl$_2$ then PhCH$_2$OH; (f) resolve with L-Z-phenylalanine.

steps by first reaction with ethyl bromoacetate to give the N-benzyl pyrrole **32** followed by hydrogenation to **31.**

A second approach, and clearly the most practical on a large scale, led to the resolved benzyl ester **33** of the core structure (Scheme 1.13). First N-acetylserine methyl ester was chlorinated with PCl$_5$ and reacted with the enamine from cylcopentanone to afford **34** albeit with loss of stereochemistry at the serine α-carbon. Acid-catalysed cyclisation gave the iminoacid **35** which was hydrogenated to the endo, cis product **36** as a racemic mixture. Treatment of this material first with thionyl chloride and then with benzyl alcohol gave the corresponding benzyl ester which was resolved with L-Z-phenylalanine to give the desired (S,S,S)-form.

The synthesis of ramipril was completed either by sequential addition of alanine and then 4-phenylbutyrate to **31** or by reaction of the resolved benzyl ester **33** with the preformed pseudo-dipeptide fragment **37** (Scheme 1.14). Thus, Z-protected alanine activated as its hydroxysuccinimde ester **38** reacted directly with **31** to give **39** as a mixture of diastereoisomers which, after chromatographic separation, were separately deprotected at both the C- and N-termini. The resulting amino acid **40**, now as a single enantiomer was reductively alkylated to give ramipril as a (RS)-mixture at the new stereogenic center. Alternatively, a Michael reaction of the benzoylacrylate **41** with alanine benzyl ester followed by reduction of the ketone gave the pseudo-dipeptide fragment **37** as a 2:1 mixture of diastereoisomers in favour of the desired SS-isomer which could be readily isolated by fractional crystallisation. This material was debenzylated and condensed with the octahydrocyclopenta[b]pyrrole core **33** and the product deprotected to give ramipril.

In order to avoid the resolution step in the synthesis of (S,S,S)-benzyl ester **33**, a modification involving an intramolecular radical cyclisation procedure proved effective (Scheme 1.15). Bromocyclopentene reacted smoothly with

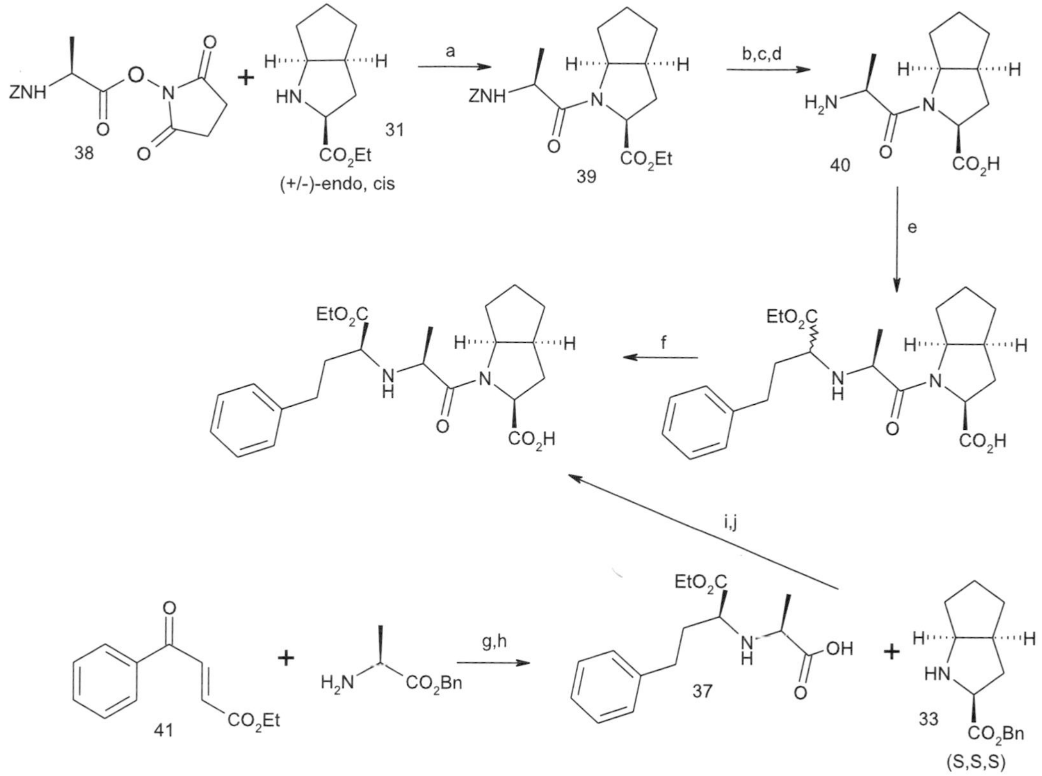

Scheme 1.14. Reagents: (a) EtOAc, rt, 20 h; (b) separate diastereomers chromatographically; (c) MeOH, 2.5N HCl, rt, 18 h; (d) MeOH, H$_2$, 10% Pd-C; (e) MeOH, Ph(CH$_2$)$_2$COCO$_2$Et, mol. sieve, 18 h, rt then NaCNBH$_3$, rt, 2 h; (f) separate diastereomers; (g) EtOH, Et$_3$N, rt, 2 h; (h) HOAc, 10% Pd-C, H$_2$; (i) K$_2$CO$_3$, H$_2$O, DCM, ethylmethylphosphinic anhydride, 3 h, rt; (j) MeOH, 10% Pd-C, H$_2$

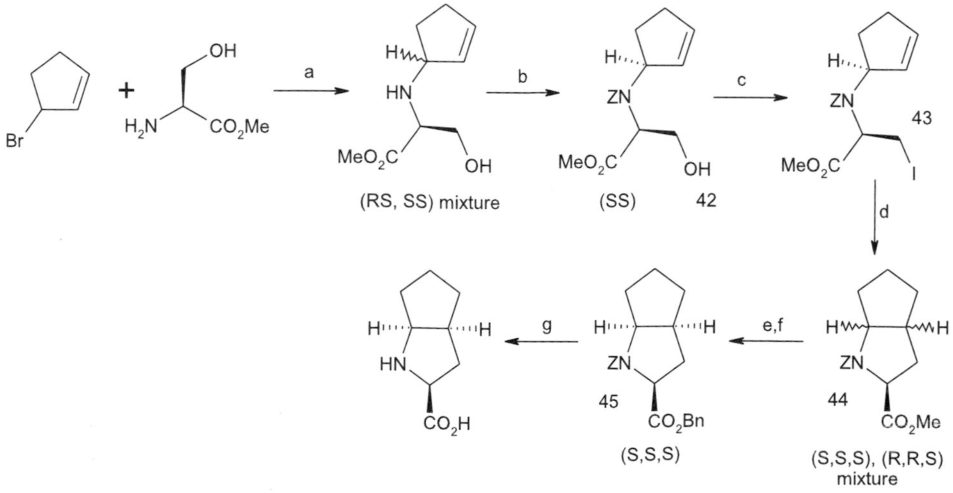

Scheme 1.15. Reagents: (a) MeCN, 0° to rt over 2h; (b) NaHCO$_3$, ClCO$_2$Bn, 0°, 1.5 h; (c) Ph$_3$P, imidazole, PhH, I$_2$, rt, 5 h; (d) AIBN, n-Bu$_3$SnH, PhH, 4 h, reflux; (e) PhCH$_2$OH, Ti(i-OPr)$_4$, 90°, 4 h; (f) separate by chromatography; (g) H$_2$, 10% Pd/C, EtOH.

L-serine methyl ester and the product treated with benzylchloroformate to give **42**. This was iodinated to **43** and the iodide subjected to a radical cyclisation using tributyltin hydride in the presence of azoisobutyronitrile (AIBN). As an aid to separation of the diastereoisomers of **44**, the material was transesterified with benzyl alcohol with titanium tetraisopropoxide as catalyst and the resulting benzyl ester **33** easily separated by column chromatography as before.

1.7 Fosinopril

Fosinoprilat **7** (R = H) represents an alternative approach to inhibitor design wherein a phosphinic acid moiety replaces the zinc ligand of earlier compounds. When given by the intravenous route (thereby bi-passing the hurdle of absorption from the gastrointestinal tract following oral adminis-tration), the compound was equally as effective as captopril in reducing the pressor (increased blood pressure) response to angiotensin II but with a longer duration of action. The drug was poorly active when given orally however and this problem was alleviated by formation of the phosphinic acid double ester (fosinopril, **7**, R = CH{CH(CH$_3$)$_2$}OCOEt acting as a prodrug of fosinoprilat. This prodrug is only slowly absorbed from the small intestine and is rapidly cleaved to active drug in the plasma.

4-Ketoproline protected as its benzyloxycarbonyl derivative **45** was used as the starting material for the C-terminus of the molecule (Scheme 1.13). Addition of phenyl magnesium bromide to **45** gave the tertiary alcohol which was dehydrated to give **46** and the double bond hydrogenated exclusively from the side remote from the carboxylic acid to give the cis disubstituted

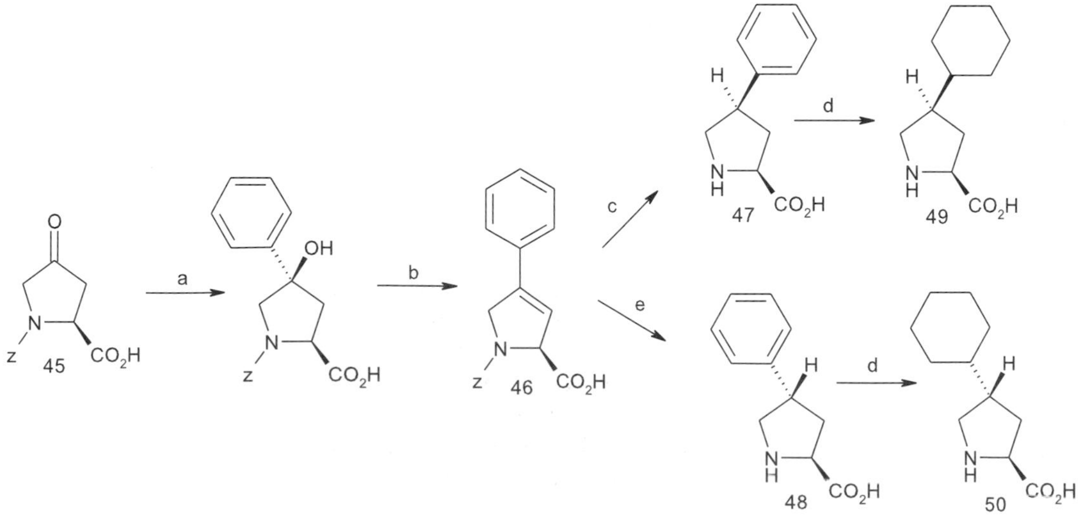

Scheme 1.13 Reagents: (a) PhMgBr, THF, 0°, then rt 18 h; (b) DCM, TFA, 18 h, rt; (c) H$_2$, Pd-C; (d) PtO$_2$, H$_2$, EtOH; (e) THF, Li, NH$_3$, −78°, 10 min.

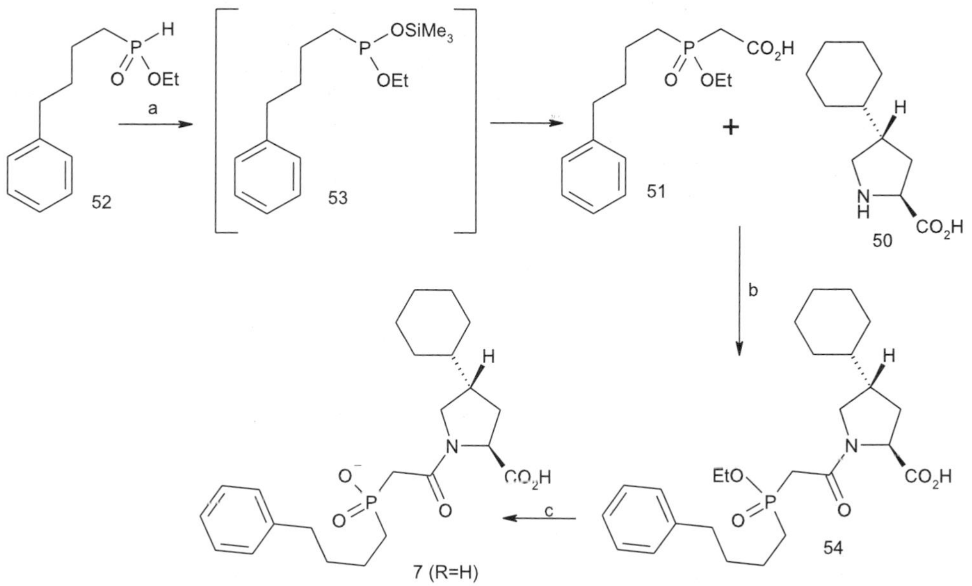

Scheme 1.14. Reagents: (a) CHCl$_3$, Me$_3$SiCl (2 equivs.), BrCH$_2$CO$_2$H, 2h, 0°; (b) CDI, MeCN, 0° then rt 48 h; (c) DCM, Me$_3$SiBr, 16 h, rt.

pyrrolidine **47** as a single enantiomer. The trans isomer **48** was also prepared from **46** but using metal – ammonia reduction followed by multiple recrystallisation to remove the minor cis isomer. In both cases, reduction of the phenyl ring using platinum catalysed hydrogenation afforded the 4-cyclohexylproline isomers **49** and **50** ready for the final coupling reaction.

The second component in the coupling reaction, the substituted phosphinic acid **51** was available from an Arbuzov reaction on the phosphonous acid ester **52** (Scheme 1.14). The reaction was carried out in the presence of trimethylsilyl chloride and proceeded via the intermediacy of the tervalent phosphonite **53**. Using carbonyldiimidazole to activate the free carboxylic acid of **51**, it was possible to prepare **54** directly without protecting the proline carboxylic acid. The free phosphinic acid **7** (R = H) was released from the ethyl ester by reaction with trimethylsilyl bromide. Surprisingly, it was found that both the cis and trans isomers had similar activity *in vivo* although it is the (2S,4S)-form which is the marketed product.

Fosinopril itself was prepared by esterifcation of the phosphinic acid at an early stage (Scheme 1.15) The intermediate **55**, having two stereogenic centers, consists of two pairs of racemates and these diastereomers were separated by fractional crystallisation from diisopropylether and the resulting racemates resolved using cinchonidine to form the chiral salt. However, the drug is sold as a mixture of isomers in the prodrug form presumably there being little difference in the rate of hydrolysis *in vivo*.

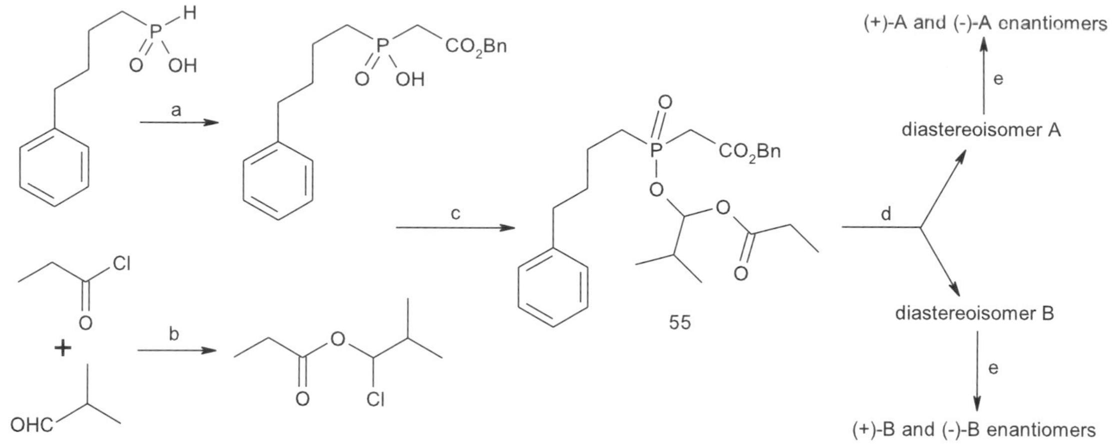

Scheme 1.15   Reagents: (a) CHCl₃, Et₃N, Me₃SICl, BrCH₂CO₂Bn, 0° then rt 5 h; (b) ZnCl₂, CH₃Cl, rt, 3 h; (c) CHCl₃, ET₃N, nBu₄NHSO₄, NaI, reflux, 20 h; (d) fractional crystallisation from (i-Pr)₂O; (e) resolve cinchonidine salt.

References

Captopril:

M. A.Ondetti, B. Rubin and D. W. Cushman, *Science*, 1977, **196**, 441.

M. A.Ondetti and D. W. Cushman, US Patent 1977, 4,046,889.

S. G. Davies, *Pure & Appl. Chem.* 1988, **60**, 13

Enalapril and Lisinopril:

A. A. Patchett et al, *Nature*, 1980, **288**, 280.

A. A. Patchett, E. E. Harris, M. J. Wyvratt and E. W. Tristram, US Patent, 1980, 4,374,829

M. J. Wyvratt et al, *J. Org. Chem.*, 1984, **49**, 2816.

M. T. Wu et al. *J. Pharm. Sci.* 1985, **74**, 352.

Quinapril:

S. Klutchko et al, *J. Medicin. Chem.*, 1986, **29**, 1953.

M. L. Hoefle and S. Klutchko, US Patent 1982, 4,344,949

Benazepril:

J. W. H. Watthey et al, *J. Medicin. Chem.*, 1985, **28**, 1511

S. K. Boyer, *Helv. Chim. Acta.*, 1988, **71**, 337

J. W. H. Watthey, US Patent, 1983, 4,410,520

Ramipril:

V. Teetz, R. Geiger, R. Henning and H. Urbach, *Arzneim. Forsch.*, 1984, **34**, 1399

H. Urbach and R. Henning, *Heterocycles*, 1989, **28**, 957

E. H. Gold, B. R. Neustadt and E. M. Smith, US Patent, 1986, 4,587,258

Fosinopril:

J. Krapcho et al, *J. Medicin. Chem.*, 1988, **31**, 1148

E. W. Petrillo et al, US Patent, 1982, 4,337,201 and 1987, 4,873,356

2 Blockade of angiotensin-II receptors

2.1 Introduction

As discussed in Chapter 1, angiotensin converting enzyme (ACE) inhibitors have become first line therapy for the control of high blood pressure and are also important in the treatment of chronic heart failure. ACE, however, is a non-specific protease and includes bradykinin, substance P and enkephalins as additional substrates. The cough associated with ACE inhibitor therapy in 5–10% of patients has been attributed to bradykinin potentiation; other points for intervention in the renin-angiotensin cascade might overcome this and other side effects. Inhibition of renin, the rate limiting enzyme in the RA cascade, has proved to be one of the most ill-fated drug discovery targets for medicinal chemists because, despite hundreds of man years of synthetic effort, there are no orally effective renin inhibitors on the market. Indeed, the success of tackling proteases as therapeutic targets is, in general, disheartening. Although many highly potent protease inhibitors are known, few have been successful in overcoming the later hurdles in the drug discovery process such as oral bioavailability, duration of action and toxicity. The most notable exception is HIV protease (Chapter 11).

Although many peptide-based antagonists of angiotensin-II receptors (the AT_1 sub-type predominates in vascular tissue) had been long known, it was not until 1982 that the first, albeit weakly active and non-selective, non-peptide inhibitors (e.g. **1**) were discovered following the chance observation of cardiovascular properties in molecules originally made as anti-inflammatory agents. This finding taken together with speculations on the active conformation of angiotensin-II (A-II) from molecular modeling eventually led to the synthesis of losartan (**2**, Scheme 2.1).

1

Losartan is as effective as enalapril in reducing blood pressure in mildly hypertensive patients. Interestingly, it has a longer duration of action than that anticipated from plasma levels of drugs suggesting an active metabolite. It is likely that the acid derived from liver induced oxidative metabolism of the parent primary alcohol dominates the therapeutic effects of losartan. Although there are numerous 'second generation' compounds progressing through clinical evaluation, only valsartan is approved to challenge the dominance of losartan.

2.2 Losartan

Two key components were required for the convergent route leading to losartan (Scheme 2.1) – the preformed imidazole **3** and a substituted biphenyl derivative **4**. Valeroamidine hydrochloride was condensed with dihydroxyacetone in liquid ammonia to afford the hydroxymethylimidazole which was chlorinated to give **3**. The 4'-bromomethyl-biphenyl **4** was prepared by the Ullmann biaryl synthesis followed by NBS bromination and this was used to alkylate the foregoing imidazole. The desired 4-chloro-5-hydroxymethylimidazole **5** was isolated as the major isomer after chromatography. The unwanted regioisomer **6** was removed by chromatography on silica since it moved significantly slower possibly because the OH group is free to interact

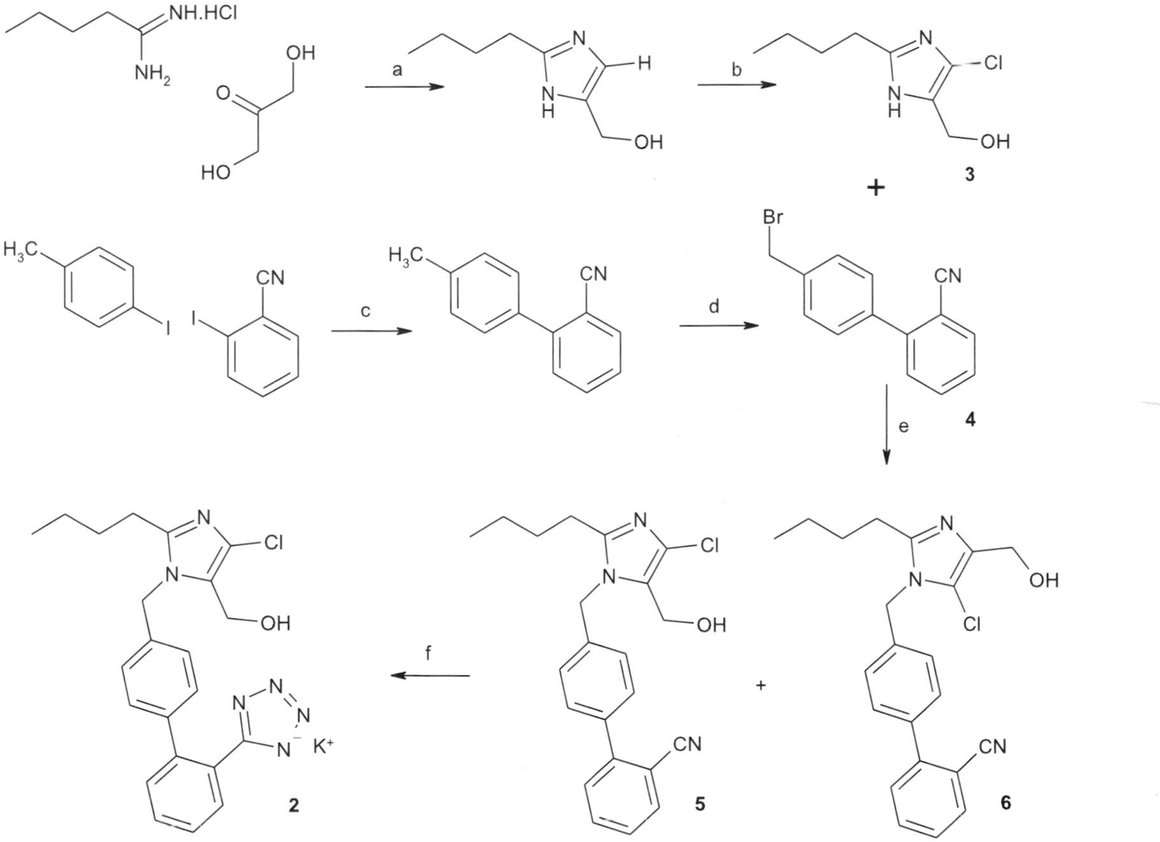

Scheme 2.1. Reagents: (a) liq. NH$_3$, 65–70°, 5 h; (b) N-chlorosuccinimide; (c) Cu powder, 210°, 2 h; (d) N-bromosuccinimide, azobisisobutyronitrile, CCl$_4$, reflux 3 h; (e) NaOMe, MeOH, 0.5 h, rt, then chromatographic separation; (f) 100°, 9 days, NaN$_3$, DMF, NH$_4$Cl then KOH.

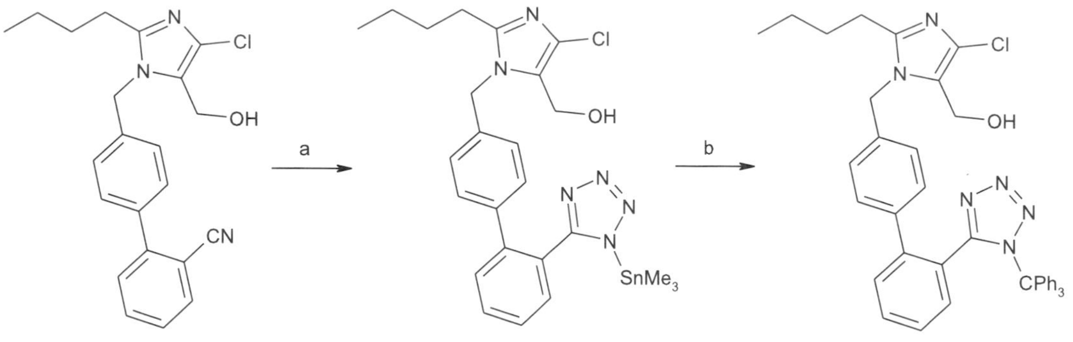

Scheme 2.2. Reagents: (a) Me$_3$SnN$_3$, xylene, 115°, 41 h; (b) 10 N NaOH, 25°, 5 min then Ph$_3$CCl, 25°, 3 h.

with the silica whereas that of the major isomer is somewhat shielded by the adjacent functionality. Absolute confirmation of structure was achieved by X-ray crystallography of **2** and related compounds within the same series.

Clearly, there are two major limitations to the route that preclude its use on the large scale: the N-alkylation step yielding variable amounts of the undesired regioisomer (often as high as 1:1) and the inefficient, albeit direct, conversion of the nitrile intermediate into the tetrazole. Direct transformation furnished a complex mixture of products after a 9 day reaction from which the tetrazole could be isolated in 32% yield. The latter problem was solved by utilising trimethyltin azide (Scheme 2.2) followed by removal of the trimethyltin group with sodium hydroxide. The product was most conveniently isolated as the triphenylmethyl derivative from which the free tetrazole was released after acid treatment.

As an alternative to the Ullmann procedure, oxazoline directed alkyation (Scheme 2.3) also allowed access to the biphenyl intermediate.

Finally, to address the regioisomer issue, the intermediacy of the imidazole aldehyde **7** in the alkylation step was studied since earlier work had suggested that the 5-carboxaldehyde should predominate. Thus, ceric ammonium nitrate oxidation of the alcohol (**3**, Scheme 2.4) afforded the aldehyde which was

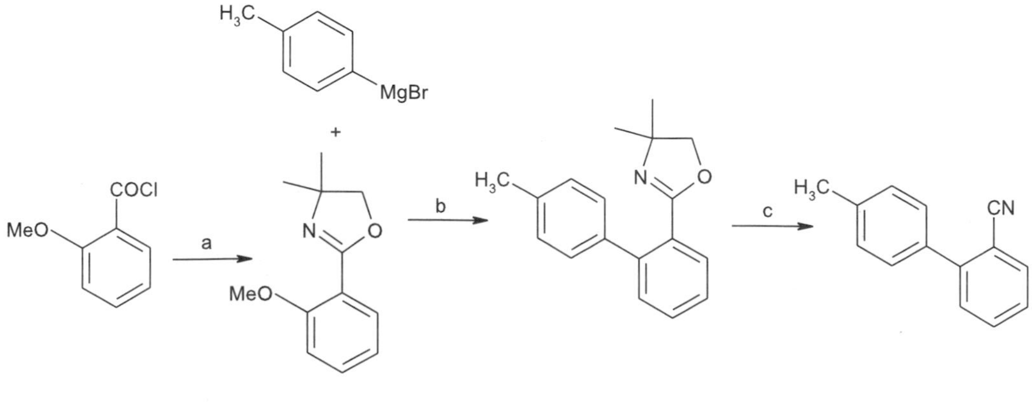

Scheme 2.3. Reagents: (a) 2-methyl-2-aminopropanol, DCM, 0° then SOCl$_2$, 25°, 1 h; (b) THF, 20°, 2 h; (c) POCl$_3$, pyridine at 15° then 100°, 3 h.

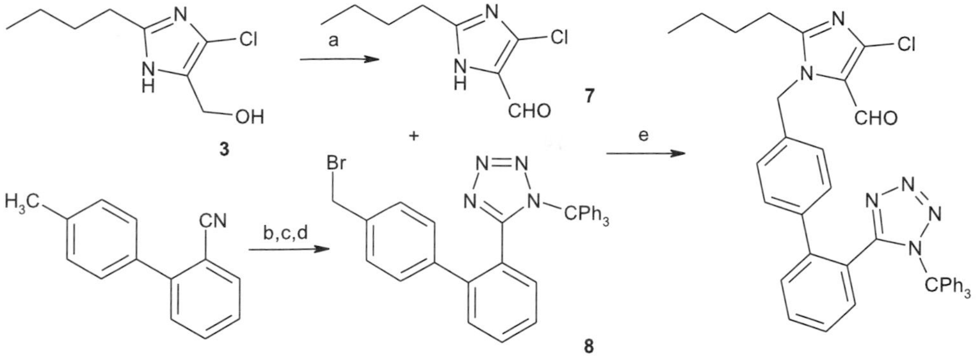

Scheme 2.4. Reagents: (a) CAN, HOAc, 30°, 3 h; (b) Bu$_3$SnCl, NaN$_3$, PhMe; (c) Ph$_3$CCl, NaOH; (d) NBS, dibenzoylperoxide, CCl$_4$, 3 h, reflux; (e) K$_2$CO$_3$, DMF, 25°, 24 h.

reacted with the preformed protected tetrazole **8** to give a 9:1 ratio of the desired N-alkylated product together with its regioisomer. Borohydride reduction and removal of the protecting group completed the synthesis.

2.3   Valsartan

Valsartan **9** (Scheme 2.5) was first approved in 1996 and, unlike losartan, it is not a prodrug with the major metabolite, 4-hydroxyvaleryl valsartan, having

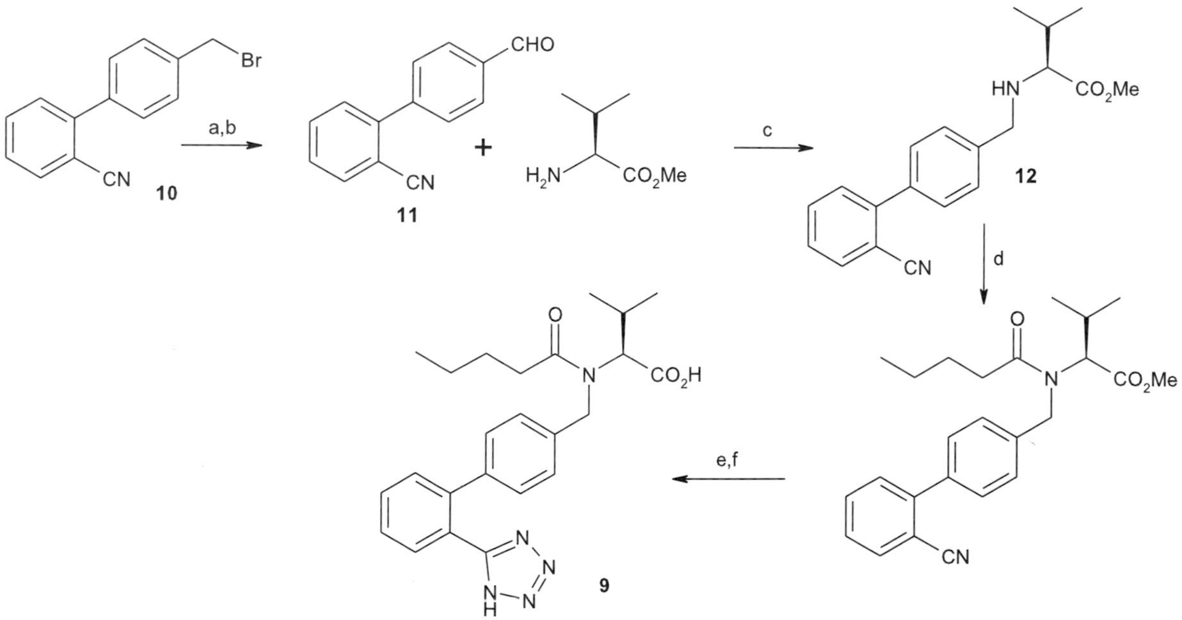

Scheme 2.5. Reagents: (a) NaOAc, HOAc, reflux 16 h; then NaOH, H$_2$O, EtOH, 16 h, reflux; (b) (COCl)$_2$, DCM, DMSO, −60°, 2 h; then Et$_3$N, −60° to rt; (c) NaBH$_4$, MeOH, THF, 5°, then rt 24 h; (d) Et$_3$N, n-valeryl chloride, DCM, 0°, then rt, 16 h; (e) Bu$_3$SnN$_3$, xylene, reflux 24 h; (f) 1N NaOH, rt, 10 h.

only one tenth of the activity of the parent. Starting with the known bromomethylbiphenyl **10**, hydrolysis to the primary alcohol via the acetate and Swern oxidation afforded the aldehyde **11**. This material was reductively aminated with L-valine methyl ester and the resulting secondary amine **12** was acylated with valeric acid chloride. As we have seen above (Scheme 2), the nitrile was converted into the tetrazole **9** using tributyltin azide under forcing conditions. The tetrazole moiety, having a pKa approximating to that of a carboxylic acid, is seen as an effective bioisotere of that group with advantages in pharmacokinetic properties.

References

Losartan:
D. J. Carini et al, *J. Medicin. Chem.*, 1991, **34**, 2525
D. J. Carini et al, US Patent, 1992, 5,138,069
R. R. Wexler et al, *J. Medicin. Chem.*, 1996, **39**, 625

Valsartan:
P. Buhlmayer et al, *Bioorg. Medicin. Chem. Lett.* 1994, **4**, 29
P. Buhlmayer et al, US Patent, 1995, 5,399,578

3   Calcium channel blockers in the treatment of angina and hypertension

3.1   Introduction

The chest pain associated with angina results from a myocardial oxygen deficiency when the oxygen supply to the heart muscle is compromised because of impeded blood supply. Interventions which reduce oxygen demand and simultaneously improve supply clearly represent an ideal therapy. It is exactly this profile that is achieved by calcium channel modulators (specifically 'blockers'); they reduce heart rate without affecting the force of contraction and have vasodilatory effects on coronary arteries, thereby reducing 'after load' on the heart. Their ability to dilate peripheral blood vessels also enables these agents to be used in hypertension.

Two distinct types of calcium channels have been implicated in regulating vascular and cardiac tone: the voltage-gated and ligand-gated ion channels. The voltage gated channels are activated by changes in electrical potential across the cell membrane; opening of such channels allows calcium to enter the cell and triggers the release of calcium from internal stores. The ligand gated channel is associated with receptors such as the adrenergic receptor which, once activated by noradrenaline for example, also cause an influx of calcium into the cell. The overall effect is to raise the resting levels of intracellular calcium from 10^{-7}M up to 10^{-5}M leading to formation of a complex of calcium with the intracellular protein calmodulin and this turn sets in motion the cascade of events that lead to muscle contraction.

The drugs reviewed in this section act predominantly on the voltage gated channels. At least three distinct groups of classes of compounds having this property have been identified as represented by nifedipine (and other dihydropyridines), diltiazam and verapamil; whether these operate at the same binding site on the calcuim channel has not been determined.

3.2   Nifedipine

Nifedipine **1** belongs to the dihydropyridine series and was the first such agent to become approved. The synthesis is implicitly simple – a classical Hantzsch condensation gives the dihydropyridine in one step from available starting materials (Scheme 3.1). Given that the dihydropyridine is symmetrical, there is no stereogenic center in the molecule although this is not the case for later drugs of this class (see below). In the course of this synthesis, it was noticed

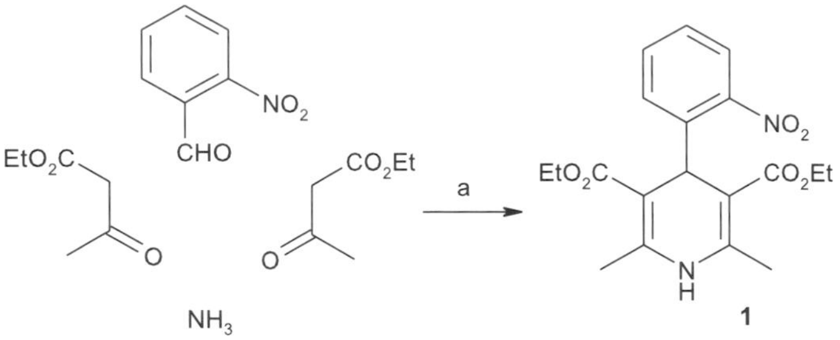

Scheme 3.1. Reagents: (a) MeOH, aq. NH$_3$, reflux 16 h.

that several products were detected by TLC and these impurities became problematic on a commercial scale. A more preferred method, albeit two stages, involves isolation of the intermediate Knoevenagel product, the benzylidine diester **2** followed by a Hantzsch reaction with ethyl 3-aminocrotonate (Scheme 3.2).

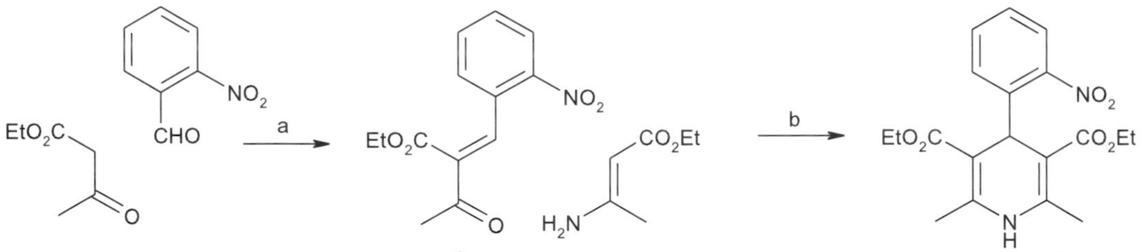

Scheme 3.2. Reagents: (a) HOAc, piperidine, 16 h; (b) EtOH, reflux, 24 h.

3.3 Amlodipine

As almost always happens with the first member of a mechanistic class, certain deficiencies became apparent following the widespread therapeutic use of nifedipine thereby giving the second generation drugs an opportunity to capture a significant share of the market. One such compound is amlodipine **3** whose success is based on a much longer duration of action in man to the extent that once daily dosing can effectively manage hypertension in most patients. Since the compound is no longer symmetrical, amlodipine exists as a mixture of two enantiomers which, once resolved, display markedly different calcium channel blocking activities with the (–)-enantiomer (as the maleate salt) being 1000-fold more active. The drug is sold as the racemate, however, and was first prepared by successive Knoevenagel and Hantzsch reactions (Scheme 3.3). The azide component **4** was available from 2-azidoethoxide and ethyl 4-chloroacetoacetate.

To achieve the synthesis of the two separated enantiomers of amlodipine it was necessary to distinguish between the two ester groups so that a diastereomeric derivative **5(a-b)** could be made with the enantiomerically pure

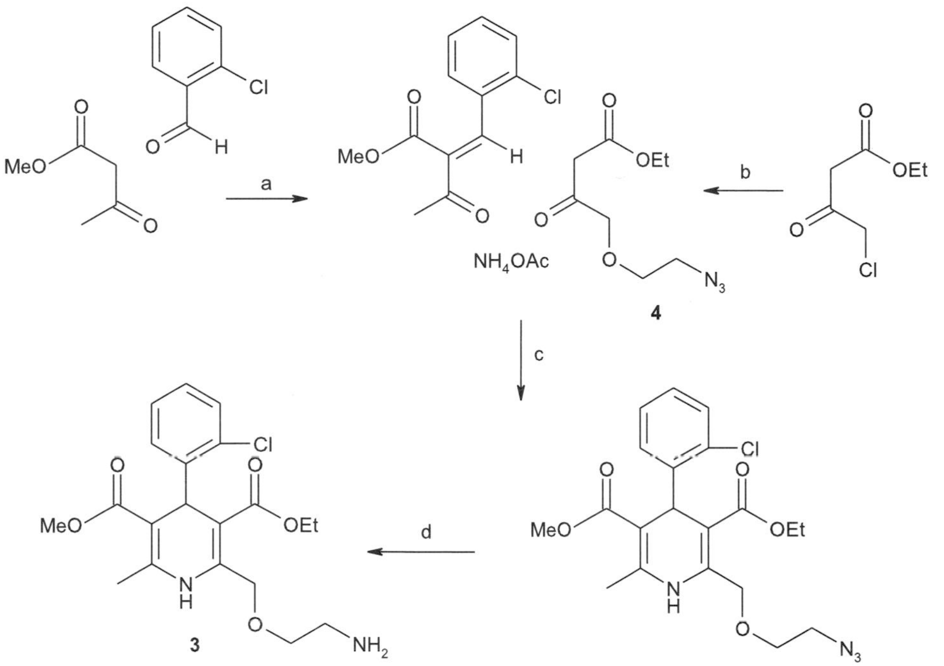

Scheme 3.3. Reagents: (a) HOAc, piperidine, 16 h; (b) 2-azidoethanol, NaH, THF, 0° then 16 rt; (c) NH₄OAc, EtOH, reflux, 2.5 h; (d) H₂, 5%-Pd on CaCO₃, EtOH.

(S)-(+)-2-phenylethanol (Scheme 3.4). For this reason, the Hantzsch condensation utilised **6** as the third component and this was derived from the acid **7** and Meldrum's acid **8** acting as a two-C source. Careful base hydrolysis of the cyanoethyl ester **9** afforded the mono-acid **10** which was activated with carbonyl di-imidazole and coupled with the homochiral alcohol. The diastereomeric diesters **5(a-b)** were separated by chromatography and individually transesterified and reduced to (−)- and (+)-amlodipine (**3a and 3b**). The absolute stereochemistry of the more active (−) enantiomer, originally incorrectly assigned, was shown to be (S) by X-ray structure analysis of **11** using (−)-(1S)-camphanic acid chloride **12** as the chiral probe (Scheme 3.5). Thus the active configuration is in line with other chiral dihydropyridines for which a structure has been determined.

3.4   Diltiazem

Diltiazem **13** originated from research into a series of benzothiazepines as novel antidepressants some members of which were shown to have potent coronary vasodilatory activity. The marketed drug is the (+)-*cis*-form **13** which was subsequently shown to be the (2S,3S)-enantiomer. It is available starting from *trans*- cinnamic acid methyl ester which was then epoxidised (Scheme 3.6) and the epoxide opened with the anion of 2-nitrothiophenol. The product **14** was resolved with cinchonidine to afford the hydroxyester **14a** as a

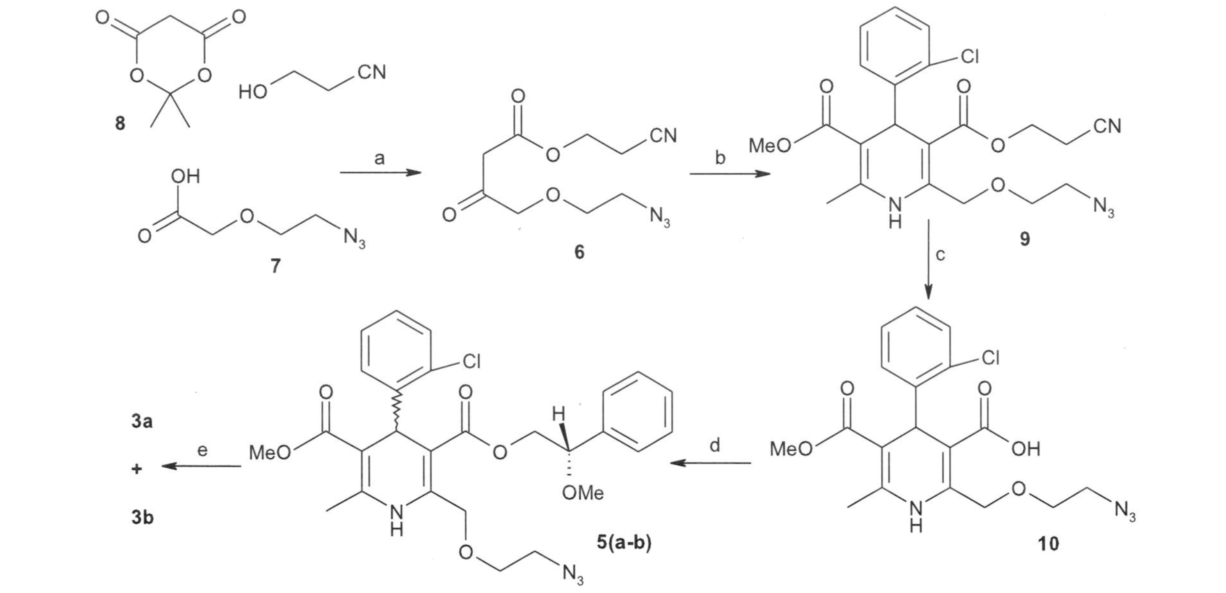

Scheme 3.4. Reagents: (a) CDI, CH$_2$Cl$_2$, py, rt 16 h; (b) 2-ClC$_6$H$_4$CHO, methyl 3-aminocrotonate, 2 h, reflux; (c) diglyme, NaOH, rt, 2 h; (d) CDI, CH$_2$Cl$_2$, S-(+)-2-methoxy2-phenylethoxide, rt, 1 h; (e) separate diastereomers then EtOH, diglyme, NaOEt, reflux,

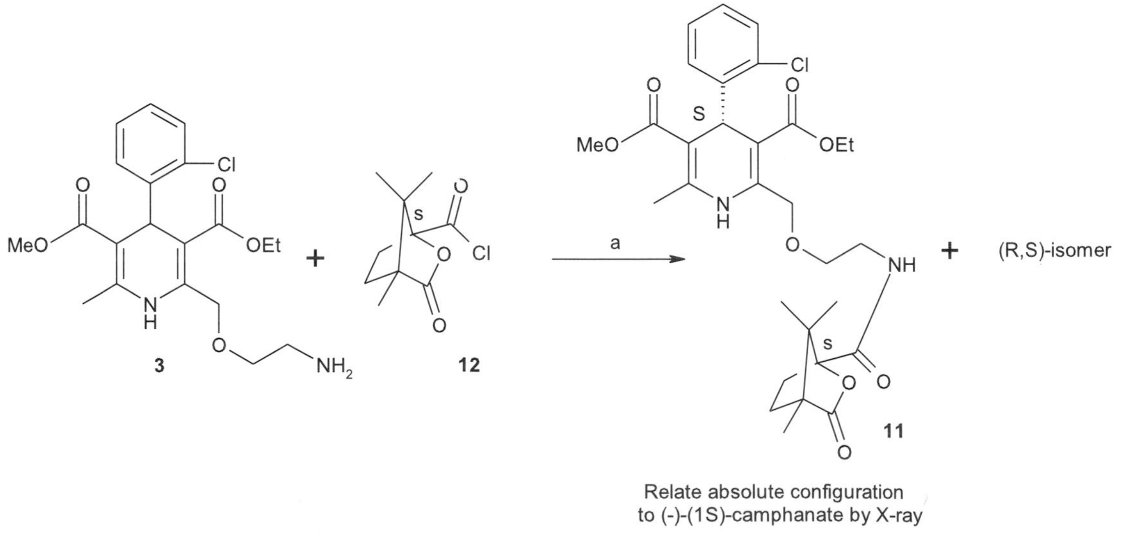

Relate absolute configuration
to (-)-(1S)-camphanate by X-ray

Scheme 3.5. Reagents: (a) CH$_2$Cl$_2$, Py, 2 h, rt, then separate on Chiralcel OD column.

single enantiomer. This was then reduced and hydrolysed to the amino acid **15** and this material cyclised to the key lactam **16.** Selective N-alkylation of the lactam using dimsyl sodium as the base followed by acetylation afforded diltiazem as the more active (+)-*cis*-isomer.

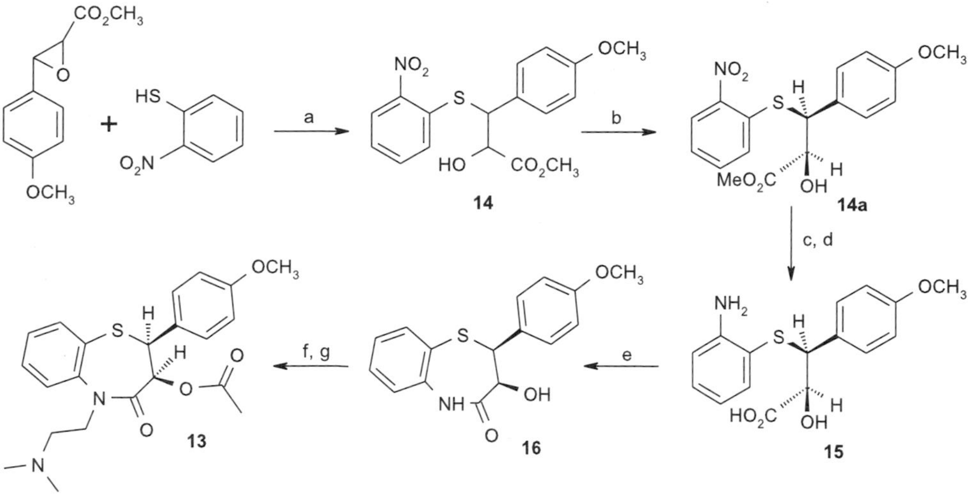

Scheme 3.6. Reagents: (a) heat together at 130°, 3 h; (b) resolve with cinchonidine; (c) 5% NaOH, EtOH, rt, 1 h; (d) 10% Pd-C, HOAc, H_2; (e) reflux xylene under Dean & Stark conditions; (f) NaH, DMSO, 1 h, rt then $(Me)_2N-(CH_2)_2-Cl$, 50°, 1.5 h; (g) Ac_2O, 5 h, 100°.

After the Sharpless procedure was established, it became possible to prepare the active enantiomer directly by substituting a *trans*-cinnamyl alcohol as the starting material (Scheme 3.7). In the first step, *trans*-4-acetoxycinnamyl alcohol was stereoselectively epoxidised and the epoxy alcohol **17** re-

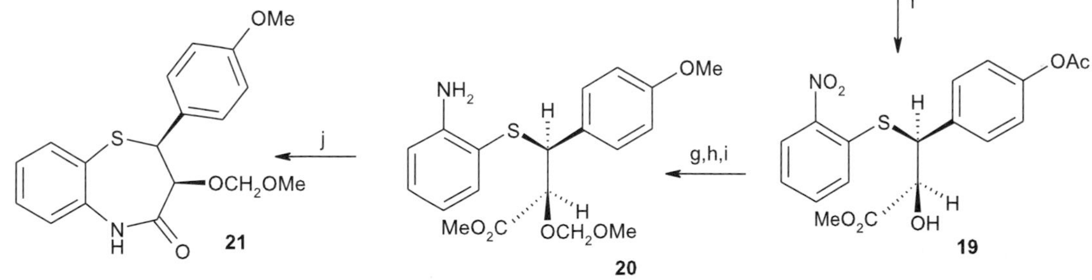

Scheme 3.7. Reagents: (a) py, Ac_2O, rt 1 h; (b) dioxane, Et_3N, $ClCO_2Et$, 30 min., 10° then $NaBH_4$, 10°, 1 h; (c) CH_2Cl_2, cumene hydroperoxide, $Ti(i-PrO)_4$, diethyl L-(+)-tartrate, 45 min., rt; (d) $NaHCO_3$, H_2O, MeCN, CCl_4, RuO_2, $NaIO_4$, 20 h, rt; then Me_2SO_4, $NaHCO_3$, DMF, 2 h, rt; (e) py.HCl, MeCN, then HCl, rt, 3 days; (f) 2-NO_2-C_6H_4SH, Et_3N, MeCN, rt, 3 days; (g) $CHCl_3$, MeCHO, P_2O_5, rt, 2.5 h; (h) THF, $BnNH_2$, 30°, 2 days then MeOH, CH_2N_2, Et_2O; (i) NH_4OH, Fe_2SO_4,, MeOH, 2 h, reflux; (j) NaOH, THF, H_2O, rt, 20 h then Et_3N, $ClCO_2Et$, THF, 1 h, rt.

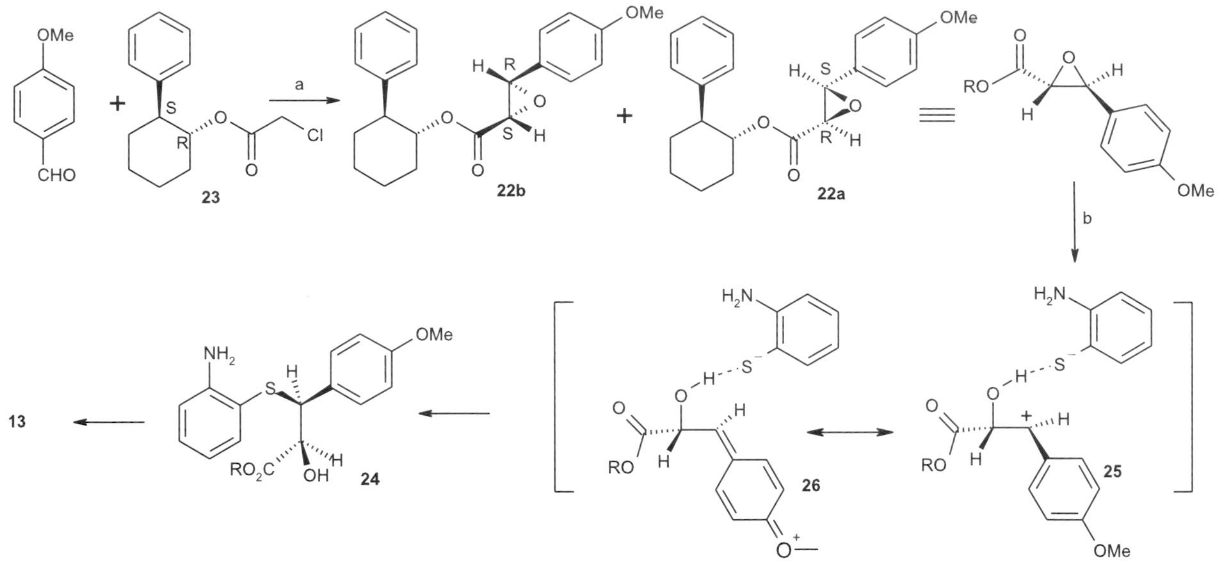

Scheme 3.8. Reagents: (a) NaH, THF, reflux, then 16 h, rt; (b) 2-aminobenzenethiol, PhMe, 16 h, reflux.

oxidised back to the carboxylic acid ester **18**. Two inversions of configuration at the 3-position (propanoic acid numbering) then followed so that the final configuration at this center was (S) in the thioether **19** and hence in diltiazem also. The process was regioselective but the reaction product was contaminated with a small amount of the (2S,3R) isomer of **19** which was removed by crystallisation. At this stage, alcohol protection using dimethoxymethane followed by deprotection of the phenol group and methylation using diazomethane afforded **20** ready for reduction of the nitro function. Cyclisation was accomplished by hydrolysis of the methyl ester and activation of the resulting acid as the mixed anhydride. Ditiazem was accessible from **21** by N-alkylation (see above), deprotection of the secondary alcohol in an exchange reaction catalysed by titanium tetrachloride with concomitant acetylation.

An alternative enantioselective synthesis exploited asymmetric induction during a Darzen's glycidic ester condensation used to prepare the glycidic ester **22a** (Scheme 3.8). The success of the approach relied on the serendipitous finding that the diastereoisomers resulting from the Darzen's reaction, wherein the two components were the chloroacetate **23** and 4-methoxybenzaldehyde, were markedly different in their respective solubilities. Additonally, it was the desired diastereomer **22a** that was the major product and it was conveniently isolated by direct crystallisation. In contrast to the outcome witnessed in the earlier synthesis (Scheme 3.7), where the thiol reaction proceeded with inversion, it was obligatory to develop conditions whereby **22a** could be converted into **24** with retention of configuration. Acid- and base-catalysed reactions gave epimerisation and inversion respectively.

On the other hand, the thermal process ensued with retention (note introduction of the 3-thiol causes a priority change at C-2) by a mechanism postulated to involve firstly protonation of the epoxide by the thiol to give an intermediate **25** stabilised by quinone methide cation **26** with the anion being held as a tight as a tight ion pair.

3.5 Verapamil

Verapamil **27** was the first calcium channel blocker to be approved in the USA although its mechanism of action was not elucidated until that of the dihydropyridines became know. Like many drugs from the 1960–1970 era, verapamil is sold as the racemate although it is the (2S)-(−)-isomer which is more active at the calcium channel. The first synthesis therefore used racemic α-isopropylveratrylnitrile and **28** as the key intermediate (Scheme 3.9); alternatively the order of alkylation may be reversed with little change in overall yield.

The separate enantiomers of **27** were first prepared (Scheme 3.10) by resolution of the acid **31** with cinchonidine and this intermediate also allowed the absolute configuration of each enantiomer to be assigned by reference to a homochiral standard (see Scheme 3.12). Although direct alkylation of phenylacetic acid methyl ester often runs the risk of double alkylation, in this instance only the mono-alkylated material **32** was isolated because of steric hindrance in the product. Under essentially the same reaction conditions but using the more activated allyl bromide, it was possible to generate the tertiary center in **33** and this could be saponified to the acid **31** as the racemic mixture. Addition of an equimolar amount of cinchonidine in methanol afforded a precipitate from which (+)-**31b** was isolated after two recrystallisations of the salt form and liberation of the free acid. The (−)-isomer was accessed from the mother liquors in a similar manner.

Scheme 3.9. Reagents: (a) NaNH$_2$, PhMe, i-PrBr, reflux, 3 h; (b) NaNH$_2$, PhMe, 5 h, reflux; (c) NaNH$_2$, PhMe, **28**, 6 h, reflux; (d) NaNH$_2$, PhMe, i-PrBr, reflux 6 h.

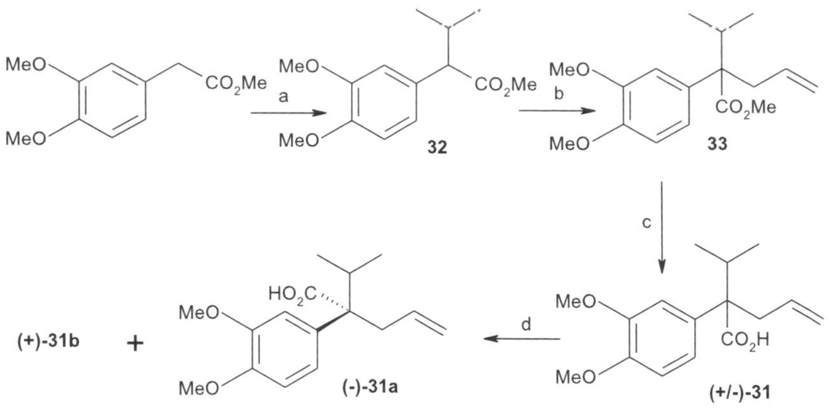

Scheme 3.10.   Reagents: (a) liq. NH$_3$, Na, Fe(NO$_2$)$_2$, i-propyl iodide; (b) liq. NH$_3$, Na, Fe(NO$_2$)$_2$, allyl bromide; distill; (c) DMSO, KOH, 48 h, 90°; (d) separate isomers with cinchonidine, fractional crystallisation.

Protection of the (−)-acid **31a** as the acid chloride (yes}!!) (Scheme 3.11) followed by hydroboration of the double bond using di-isoamylborane gave the intermediate borane which was intercepted first by liquid ammonia and then H$_2$O$_2$ to yield the hydroxy amide **34**. Using trichloro-2-benzodioxa-1,3-phosphole, it was possible to dehydrate the amide function in **34** and simultaneously chlorinate the primary alcohol to prepare **35** directly and this intermediate was then treated with N-methylhomoveratrylamine to afford (−)-verapamil.

In order to assign the absolute configuration of each enantiomer of verapamil, the resolved intermediate **31b** was transformed into a known reference using chemistry that was free from stereochemical ambiguity. Thus the (+)-acid **31b** was converted into (+)-2-methyl-2-isopropylsuccinic acid **36** itself a degradation product of naturally occurring terpenes and which had

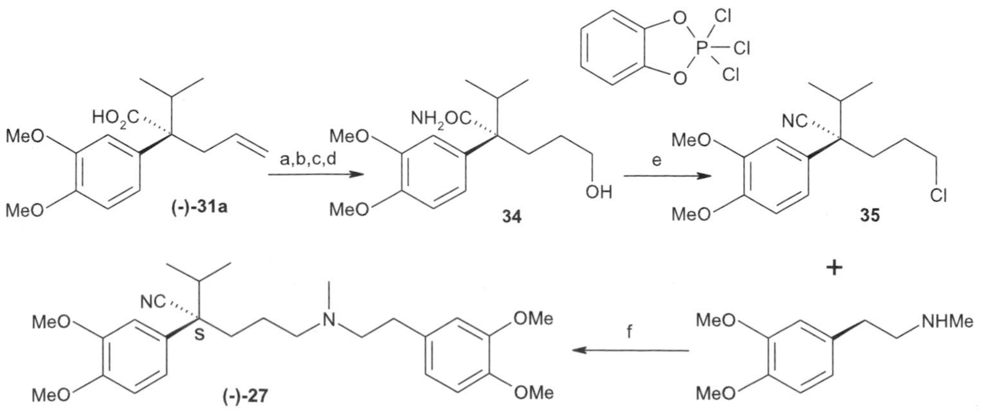

Scheme 3.11.   Reagents: (a) PCl$_5$, , Et$_2$O, 16 h, rt, then azeotrope with toluene; (b) di-isoamylborane, THF, 1 h, 20°; (c) cool to −10°, liq. NH$_3$, 1h; (d) H$_2$O$_2$, 1 h 40°; (e) trichlorophosphole, 100°, 1 h (no solvent); (f) neat mixture, 130°, 4 h.

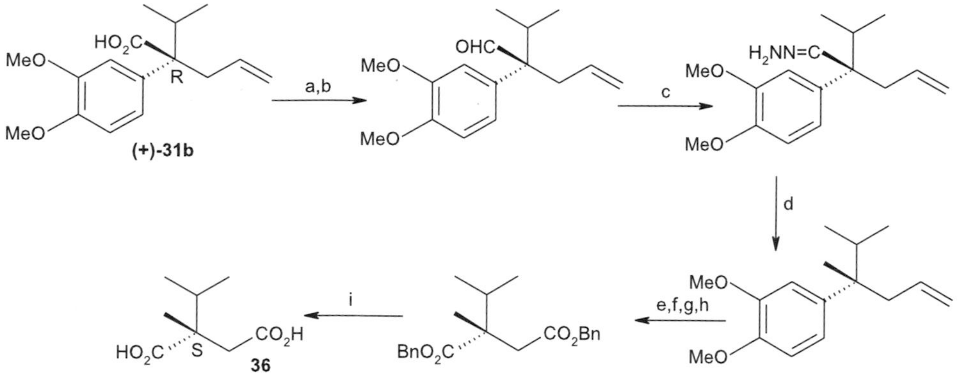

Scheme 3.12. Reagents: (a) LiAlH$_4$, THF; (b) py, CrO$_3$, 25°, 2h; (c) EtOH, NH$_2$NH$_2$, reflux, 2 h; (d) DMSO, ethylene glycol, 100°, 7 h; (e) HOAc, O$_3$; (f) H$_2$O$_2$; (g) NaOH, KMnO$_4$, H$_2$O, 45 min; (h) Et$_2$O, PhCHN$_2$, chromatography; (i) H$_2$, Pd-C, EtOH.

previously been assigned the (S)-configuration (Scheme 3.12). Bearing in mind the change in (Prelog) priority during the degradation to the 2,2-dialkylsuccinate without stereochemical inversion, it may be concluded therefore that the more active isomer of verapamil, that is the (–)-isomer, has S-configuration.

An unequivocal stereoselective synthesis of (2S)-(–)-verapamil started from the commercially available (2S)-(+)-1,2-propanediol (Scheme 3.13) the stereogenic centre bearing the methyl group which allowed stereochemical induction at the adjacent center in the key intermediate **37**. After protection and activation, the resulting mesylate **38** was displaced with the dianion of

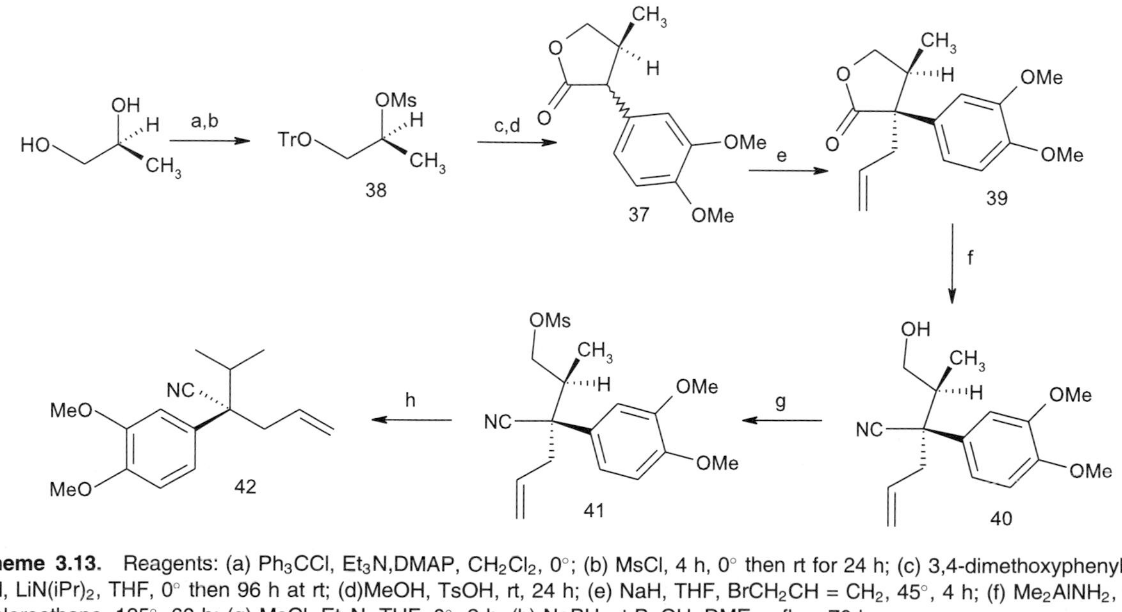

Scheme 3.13. Reagents: (a) Ph$_3$CCl, Et$_3$N,DMAP, CH$_2$Cl$_2$, 0°; (b) MsCl, 4 h, 0° then rt for 24 h; (c) 3,4-dimethoxyphenylacetic acid, LiN(iPr)$_2$, THF, 0° then 96 h at rt; (d)MeOH, TsOH, rt, 24 h; (e) NaH, THF, BrCH$_2$CH = CH$_2$, 45°, 4 h; (f) Me$_2$AlNH$_2$, 1,1,2-trichloroethane, 125°, 60 h; (g) MsCl, Et$_3$N, THF, 0°, 2 h; (h) NaBH$_4$, t-BuOH, DME, reflux, 70 h.

3,4-dimethoxyphenylacetate in an S_N2 reaction causing complete stereo-chemical inversion at the stereogenic center. Acidic deprotection afforded the lactone **37** as a mixture of diastereomers which, as the enolate, was smoothly alkylated anti to the methyl group to give **39** as the sole product in >95% ee. Noting that the $CHCH_3$ and CH_2OH moieties together are destined to become the isopropyl group in the final product and the lactone is to be converted into the nitrile, the next steps employed dimethylaluminium amide (to give **40**) followed by borohydride reduction of the mesylate **41** to give **42**. To avoid reduction of the nitrile group with borane generated in the borohydride reaction, it was necesssary to use t-butanol to destroy the borane as it was generated. With the substituents at the stereogenic center now in place, all that remained was to elaborate the allyl side chain in a manner already described above (Scheme 3.11) except that the alcohol was activated as the mesylate rather than the chloride **35**.

References

Nifedipine:
F. Bossert and W. Vater, US Patent, 1969, 3,485,847

Amlodipine:
J. E. Arrowsmith et al, *J. Medicin. Chem.*, 1986, **29**, 1696
S. Goldmann, J. Stoltefuss and L. Born, *J. Medicin. Chem.*, 1992, **35**, 3341
S. F. Campbell et al, US Patent, 1986, 4,572,909

Diltiazam
H. Kugita et al, *Chem. Pharm. Bull. (Japan)*, 1971, **19**, 595
H. Kugita et al, US Patent, 1971, 3562257
A. Schwartz et al, *J. Org. Chem.*, 1992, **57**, 851
O. Miyata et al, *Tetrahedron*, 1997, **53**, 2421
E. N. Jacobsen et al, *Tetrahedron*, 1994, **50**, 4323

Verapamil:
F. Dengel, US Patent, 1966, 3,261,859
H. Ramuz, *Helv. Chim. Acta.*, 1975, **58**, 2050
L. J. Theodore and W. L. Nelson, *J. Org. Chem.*, 1987, **52**, 1309
V. Cannata, G. Tameriani and G. Zagnoni, US Patent, 1992, 5,097,058

4 Antagonists of histamine receptors as anti-ulcer remedies

4.1 Introduction

Long before the details were known of receptor superfamilies, classification and activation mechanisms, receptors were distinguished by classical pharmacological methods. As an example, histamine was known to stimulate smooth muscle from various organs such as the gastrointestinal tract and the bronchi of the lung and, furthermore, these actions were specifically blocked by the drug mepyramine **1**. However, there are also other actions of histamine which are mepyramine-insensitive and these include the stimulation of acid secretion in the stomach. To account for such observations, the receptors were assumed to be different and were labeled H_1 and H_2 respectively. Now that the gene (and hence protein) sequences have been to determined, these receptors are known to belong to a large gene superfamily of G-protein coupled receptors (GPCR's) thus called because agonist occupation (here the agonist is histamine) of the receptor leads to activation of an associated intracellular protein (the 'G-protein'). In this way, an extracellular event (histamine binding to a cell surface receptor) may be transformed into an intracellular signal which forms the basis of the final physiological response – release of H^+ from the parietal cells in the stomach. Since it was known that the excess acid is responsible for the pain associated with peptic ulcers, it was believed that specific H_2 antagonists would both give symptomatic relief to patients and also improve the chances of healing the lesion. These ideas gave rise to the most successful class of drugs to emerge out of research in the 1970's.

With the hypothesis that antagonists may have some structural similarities to the endogenous ligand, most early studies focussed on making progressive changes to histamine. Although this proved to be a successful strategy, it also generated a dogma that the imidazole ring was crucial for activity and persuaded the original research team that its substitution by other heterocycles such as furan was inappropriate. In doing so, the most commercially successful drug ever launched was missed and handed to the competition. It is also of interest to note that this strategy does not always prove successful – note that antagonists for the related H_1 receptor are more structurally diverse (see **1** and Chapter 8) and offer only a distant resemblance to histamine. The prototypic H_2 antagonist, metiamide **2**, whilst effective in reducing gastric acid secretion in patients, had undesirable side effects possibly associated with the thiourea group. From a systematic search for isoelectronic groups, the cyanoguanidine functionality of cimetidine **3** was discovered. However, within three years, ranitidine **4** had emerged as the market leader, at its peak selling over $2 billion a year with famotidine **5** also taking a major slice of the market.

4.2   Cimetidine

Cimetidine was first reported in 1976 and its synthesis started with ethyl ester **6** (Scheme 4.1) which was originally reduced with $LiAlH_4$ to give the alcohol

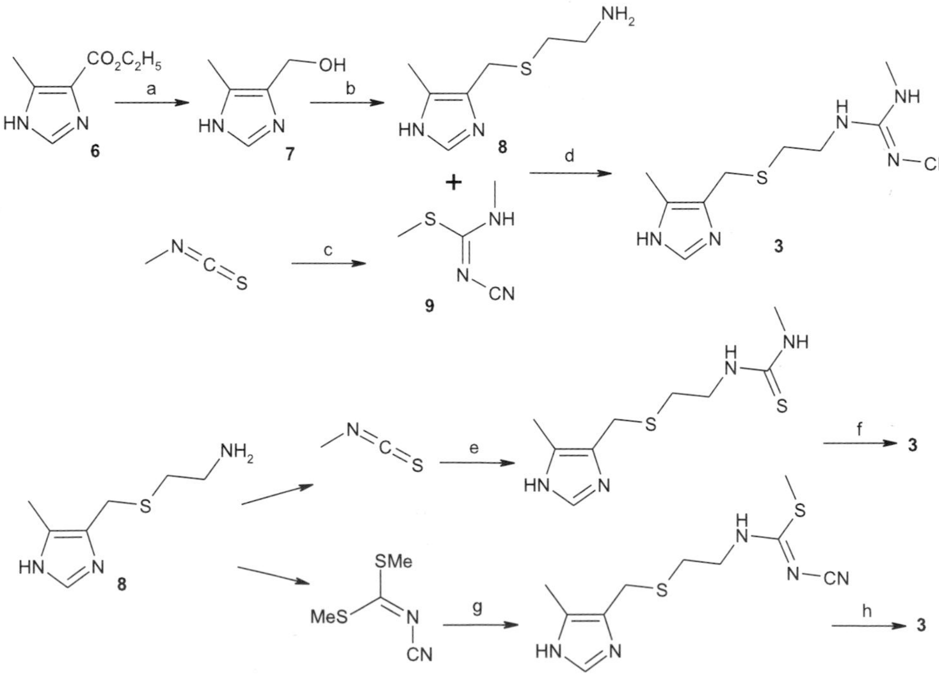

Scheme 4.1.   Reagents: (a) liq NH_3, Na, t-BuOH, then MeOH, NH_4Cl; (b) $HS(CH_2)_2NH_2.HCl$, 48% aq. HBr, reflux, 18 h; (c) NaNHCN, EtOH, 18 h, rt then MeI; (d) MeCN, reflux, 48 h.; (e) K_2CO_3, H_2O then MeN = C = S, reflux, 30 min.; (f) PbNCN, MeCN, DMF, reflux, 24 h; (g) EtOH, rt, 16 h; (h) $MeNH_2$, EtOH, rt, 2.5 h.

Scheme 4.2. Reagents: (a) H_2O, K_2CO_3, then add EtOAc and 10, 4 h, 5°; (b) i-PrOH, EtOAc, $MeNH_2$, 16 h, rt.

7; this procedure is expensive and inconvenient on a large scale. Accordingly, sodium and liquid ammonia with methanol as a proton source became the method of choice. Interestingly, over reduction to give Birch intermediates was not observed although in closely related systems this is indeed a problem. Reaction conditions had to be carefully controlled to minimise formation of the corresponding acid (after workup). It was shown that 'inverse addition' of the ester to sodium in ammonia so that the sodium was always present in high concentration achieved this objective. Acid catalysed displacement of the primary alcohol with the hydrochloride salt of cysteamine afforded the intermediate primary amine **8** which reacted under forcing conditions with N-cyano-N'-S-dimethylisothiourea **9**. The cyanoguanidine resulting from an addition-elimination sequence was isolated by chromatography followed by crystallisation from acetonitrile. The isothiourea **9** was available from methyl isocyanate by reaction first with sodium cyanamide and then methyl iodide. Several alternative methods were used to generate the cyanoguanidine as indicated but the more convergent approach proved most convenient.

Eventually, an alternative synthon for N-cyanoguanidine was developed to produce a convenient and low cost alternative final step which required diphenylcyanocarbonimidate **10** (Scheme 4.2; available from cyanamide and dichlorodiphenoxymethane) as the starting material. Two successive displacements which proceed with no gaseous, toxic side products (HSMe), gave cimetidine in high yield.

During a mechanistic study on the reactivity of 2-(N-cyanoimino)thiazolidine derivatives (e.g. **12**, Scheme 4.3) towards nucleophiles, an unexpected N to S alkyl transfer reaction occurred that was subsequently utilised in a new synthesis of cimetidine. Thus 4-chloromethyl-5-methylimidazole (as its hydrochloride) was treated with NaH and then with the sodium salt of 2-(N-cyanoimino)thiazolidine **11** to afford **12** which was then reacted with methylamine. Of the two electrophilic sites, at C(2) and the nitrile C-atom, reaction occurred exclusively at the former to give **13**. This material, when exposed to NaH in DMF, rearranged smoothly, possibly by intramolecular alkyl transfer (**14**), to give cimetidine in high yield.

4.3 Ranitidine

A Mannich reaction on 2-hydroxymethylfuran generated the first key intermediate towards ranitidine (Scheme 4.4). Acid catalysed displacement

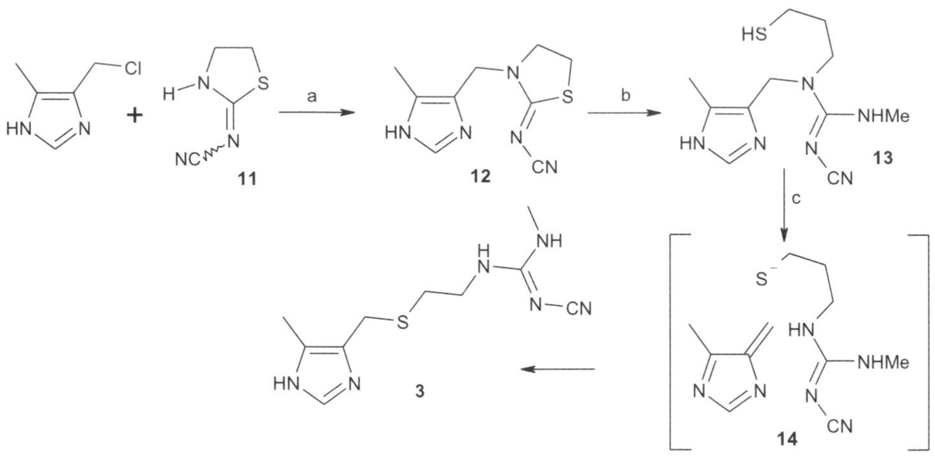

Scheme 4.3.   Reagents: (a) NaH, DMF; (b) MeNH$_2$, EtOH, sealed tube at 60°, 5 h; (c) NaH, DMF, 85°, 3 h.

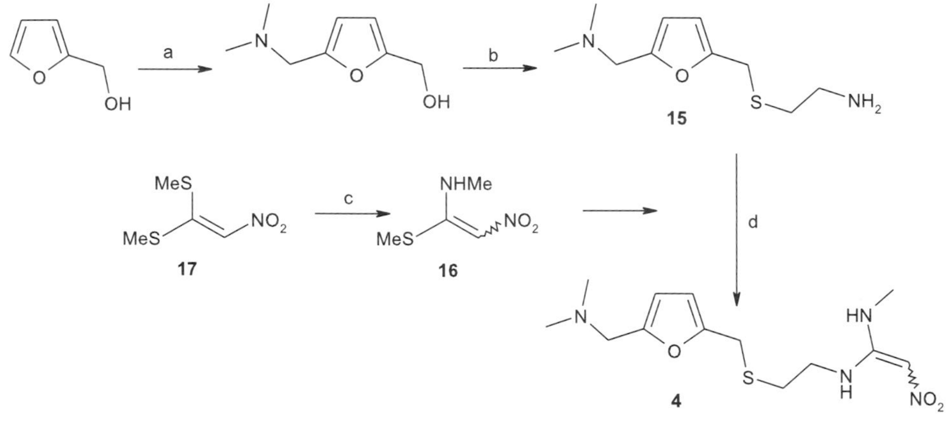

Scheme 4.4.   Reagents: (a) MeNH$_2$.HCl, aq. CH$_2$O, 0°, 3 h; (b) HS(CH$_2$)$_2$NH$_2$.HCl, conc. HCl, 18 h, 0°; (c) MeNH$_2$, EtOH, DCE, 70°, 5.5 h; (d) add **16**, 120°, 30 mins.

of the primary alcohol with cysteamine (as its hydrochloride) afforded **15** which was treated with N-methyl-1-methylthio-2-nitroetheneamine **16** in a Michael addition-elimination reaction to give ranitidine directly. The Michael acceptor was itself available from the bismethylthio nitroolefin **17** by addition of methylamine. Alternatively, the reaction order could be reversed so that the bisthiomethyl compound could be reacted with **15** and then with methylamine.

Perhaps the shortest route to ranitidine involves a two step sequence starting from the Mannich product, 5-dimethylaminomethylfurfural alcohol (Scheme 4.5). The process was based on the finding that 2-nitroethylenethia-zolidine **18**, when exposed to a primary amine such as methylamine, exists in equilibrium with the addition product **19** and this may be intercepted with an electrophile such as the chloromethylfuran **20** thereby driving the reaction to

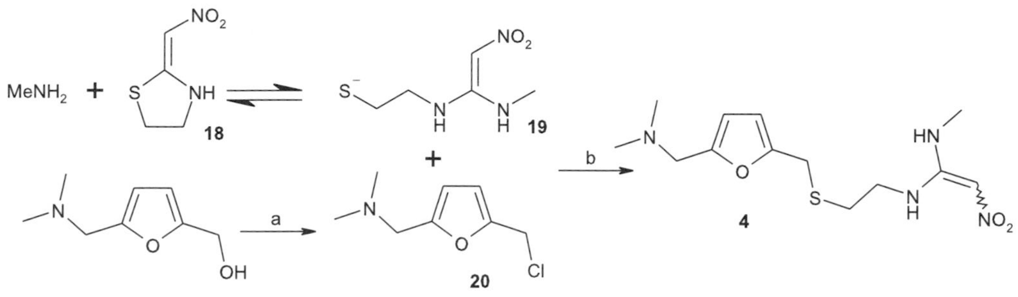

Scheme 4.5. Reagents: (a) DCM, 0°, 0.5 h; (c) IPA, MeNH$_2$, 18, 40°, 4.5 h.

completion. The intermediate **19** is unstable, cannot be isolated but, in the presence of air, readily forms the corresponding symmetrical disulphide.

4.4 Famotidine

A series of functional group modifications characterise the synthesis of famotidine. The aminothiazole **21** used as the initial building block in the synthetic route (Scheme 4.6) was readily available in one step from thiourea and dichloromethylketone. Whilst one thiourea unit is captured in the aminothiazole ring, the second unit acts as a leaving group in the following reaction – incorporation of the propionitrile group to afford **22**. Next it was necessary to convert the 2-amino group into the guanidine entity of the final drug and this was achieved by reaction first with benzoylisothiocyanate and

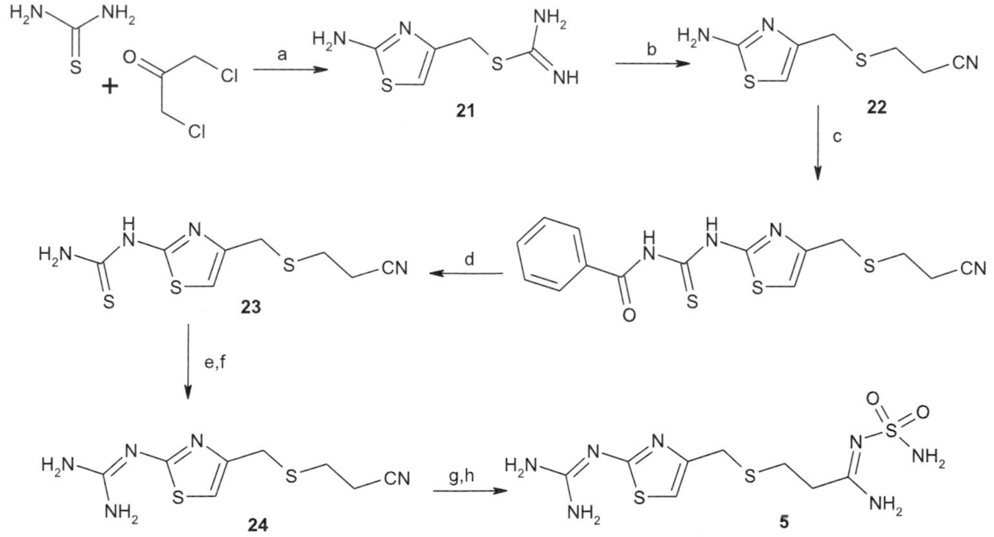

Scheme 4.6. Reagents: (a); (b) EtOH, H$_2$O, Cl(CH$_2$)$_2$CN, NaOH, 10° to rt, 2 h; (c) Me$_2$CO, PhCONCS, reflux, 5 h; (d) Me$_2$CO, H$_2$O, K$_2$CO$_3$, 50°, 5 h; (e) MeI, EtOH, reflux, 1 h; (f) MeOH, NH$_3$, NH$_4$Cl, sealed tube 85°, 15 h; (g) MeOH, CHCl$_3$, HCl gas, 0°, 4 h; (h) MeOH, NH$_2$SO$_2$NH$_2$, reflux, 3 h.

the benzoyl group removed by mild base hydrolysis. The newly formed thiourea **23** was activated by methylation and converted into the guanidine **24** by reaction with ammonia under pressure. To complete the synthesis, the nitrile function was transformed into the corresponding imino ether and this intermediate treated with sulphamide.

References

Cimetidine:
G. J. Durant, J. C. Emmett and C. R. Ganellin, US Patent, 1976, 3,950,333
G. J. Durant et al., *J. Medicin. Chem.*, 1977, **20**, 901.
R. Ganellin, *J. Medicin. Chem.*, 1981, **24**, 913.

Ranitidine:
B. J. Price, J. W. Clitherow and J. Bradshaw, US Patent, 1978, 4,128,658

Famotidine:
Y. Hirata et al., US Patent, 1981, 4,283,408.
Yanagisawa, Y. Hirata and Y. Ishii, *J. Medicin. Chem.*, 1987, **30,** 1787.

5 Proton pump inhibitors as gastric acid secretion inhibitors

5.1 Introduction

Gastric acid has long been identified as a major factor which contributes to peptic ulcer disease and its secretion is stimulated by at least three distinct mechanisms: histamine, acetylcholine and gastrin. Pharmacological inhibition of all three agents has proven effective in reducing acid secretion although inhibition of histamine acting at H_2 receptors has gained more widespread therapeutic acceptance (see Chapter 1). In 1977, however, a new class of antisecretory agent was discovered which suppresses gastric acid secretion by specific inhibition of the H^+/K^+ ATPase enzyme system at the secretory surface of the gastric parietal cell. Because this enzyme system is regarded as the acid (proton) pump within the gastric mucosa, omeprazole has been characterized as a gastric acid-pump inhibitor, in that it blocks the final step of acid production and thus curbs acid production by all stimulatory mechanisms. Recent clinical experience has indicated that this class of drug, when used in combination with clarithromycin, is an effective treatment of *Helicobacterium pylori* infections which often accompany duodenal ulcers. *H. pylori* eradication has been shown to reduce the risk of recurrence of duodenal ulcers.

5.2 The 'Omeprazole cycle'

Omeprazole, **1**, was the first proton pump inhibitor to be approved (in 1988) but its mechanism of action is not entirely straightforward and is dependent on a subtle balance of chemical reactivity, controlled by the nature of substituents in the pyridine ring, and enzymatic transformation under the physiological conditions prevailing in the gut and liver. In attempts to improve the potency of the initial lead sulphoxide, **2**, it was found that electron withdrawing substituents in the pyridine ring which increase the pKa of the pyridine nitrogen also increased activity. Thus methyl in the 3- and/or 5-position and a methoxy group in the 4-position strongly raised the basicity of the pyridine ring although, paradoxically, the substitution pattern of omeprazole acts to force the methoxy group out of the the plane of the ring thereby markedly reducing that substituent's effect.

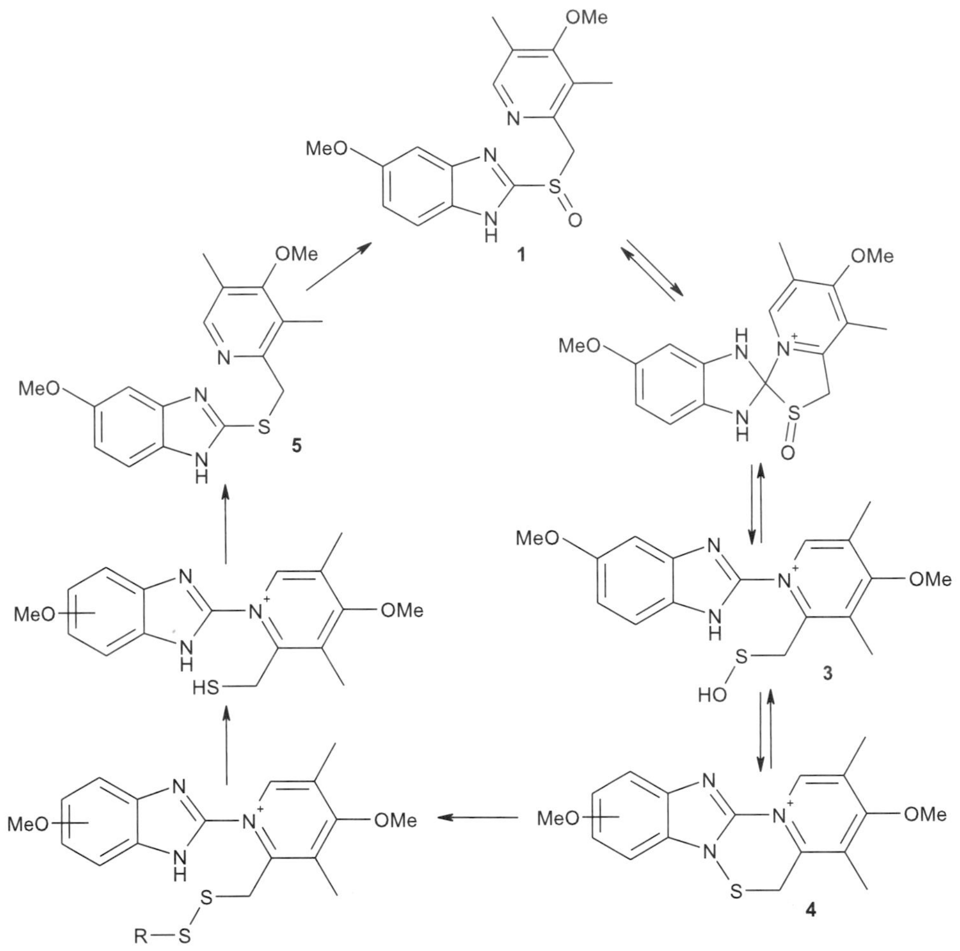

Scheme 5.1. The 'omeprazole cycle'. RSH is provided by the H⁺/K⁺ ATPase active site.

Whilst there have been several hypotheses to explain the mechanism by which omeprazole exerts its action, convincing experimental evidence has confirmed the 'omeprazole cycle' (Scheme 5.1). At the physiological pH of blood (7.4) and within the cytosol of the parietal cell, omeprazole (pKa = 4) exists as the neutral species but becomes protonated in the highly acidic environment within the sceretory cannaliculus and becomes trapped there. It is here that the molecule undergoes acid-catalysed transformation (Scheme 5.1) whereby nucleophilic attack of the pyridine-N followed by reversible ring opening to the sulphenic acid **3** and subsequent dehydration affords the pyridinium suphenamide **4**. This reactive intermediate may then be intercepted by a sulphydryl group at the active site of the ATPase – a process which may be blocked by mercaptoethanol – resulting in deactivation of the enzyme over an extended period of time. Note that this intermediate exists as a regioisomeric mixture but that distinction is lost again upon ring opening. Slow reactivation *in vivo* may result from disulphide exchange with endogenous mercaptans (such as cysteine and glutathione) and the released product **5** reconverted into the parent omeprazole by oxidative enzymes in the liver.

5.3 Synthesis of omeprazole

The process by which omprazole is made remains sketchy since it has not be fully detailed in the scientific literature. 3,5-Dimethylpyridine was first methylated with methyl lithium and then oxidised to the N-oxide **6** (Scheme 5.2) to facilitate the nitration step and remote functionalisation of the pyridine 2-methyl group. Although apparently circuitous, activation of the pyridine ring by nitration followed by displacement of the nitro group by methoxide proceeded smoothly. Functionalisation of the 2-methyl group was achieved by acetylation of the N-oxide **7** with acetic anhydride with subsequent rearrangement of the acetoxy group to give **8**. This material was hydrolysed and the resulting hydroxymethyl group activated by chlorination.

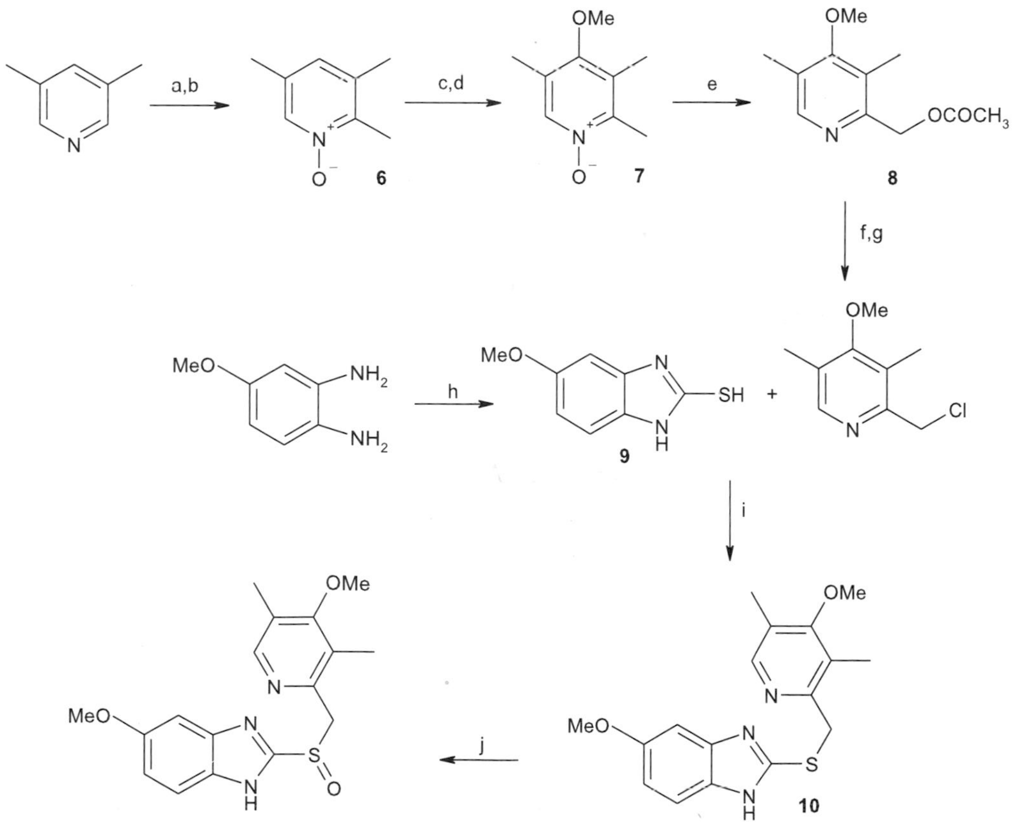

Scheme 5.2. Reagents: (a) MeLi, THF; (b) aq. H_2O_2, HOAc, 24 h, 90°; (c) HNO_3, H_2SO_4, heat; (d) MeOH, MeO⁻; (e) Ac_2O, 110°; (f) MeOH, aq. NaOH, reflux; (g) DCM, 0°, $SOCl_2$; (h) EtOC(=S)S⁻K+; (i) H_2O, EtOH, NaOH, reflux 2 h; (j) mcpba, $CHCl_3$, 5°; 10 min, pH 8.6.

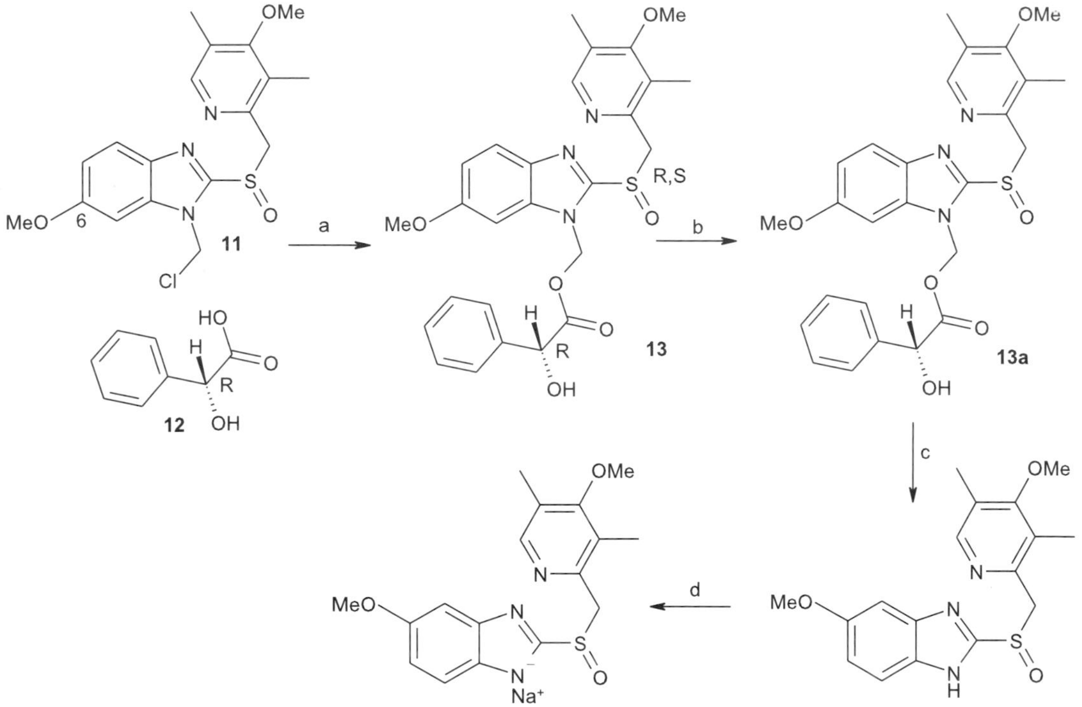

Scheme 5.3. Reagents: (a) H$_2$O, n-Bu$_4$N.HSO$_4$, NaOH then add **11** and reflux CHCl$_3$; (b) reverse phase HPLC, isolate faster running band; (c) MeOH, NaOH, 10 min, rt then HCO$_2$Me; (d) NaOH, H$_2$O, PhMe, butanone.

2-Thio-5-methoxybenzimidazole **9** (from the corresponding phenylenedia-mine) was reacted with the chloromethyl intermediate in the presence of sodium hydroxide to afford **10** which was oxidised under mild conditions with mcpba with careful control of pH by judicious titration of KHCO$_3$ to maintain a pH of 8.6.

Omeprazole is a racemic mixture having a stereogenic center at the sulphoxide S atom. Surprisingly, given the acidic nature of the adjacent methylene group, it was found that the two enantiomers are stable towards epimerisation under neutral and basic conditions making it possible to prepare and isolate sodium and magnesium salts of the separated isomers and study their properties *in vivo*. Separation of the racemic mixture involved reaction of 1-chloromethyl derivative **11** of omeprazole with either (R)- or (S)-mandelic acid **12** (Scheme 5.3) to give **13**. In either case, it proved convenient to isolate the (pure) more hydrophilic isomer after reverse phase HPLC so that (R)-mandelic acid eventually yielded (–)-omeprazole whilst (S)-mandelic acid gave the (+)-isomer. Thus solvolysis of each separated diastereomer with sodium hydroxide and neutralisation with methylformate gave (–)- and (+)-omeprazole which were isolated as the sodium salts each having an

Scheme 5.4. Reagents: (a) HOAc, pyrrolidine, PhH, azeotrope; (b) (COCl)$_2$, CHCl$_3$, reflux 25 min. then (care!) MeOH, reflux, 10 min; (c) NaBH$_4$, MeOH, 45 min., rt; (d) 32% aq. NH$_3$, 60°, 48 h; (e) POCl$_3$, reflux 1 h; (f) THF, 2N NaOH, 40°, 18 h.

enantiomeric excess (ee) greater than 99.8% but with opposite specific rotation.

In order to avoid the intermediacy of pyridine-N-oxides, long recognised as toxic, a completely different approach used the pyrone **13** (Scheme 5.4) as the key intermediate. It was envisaged that the ring O atom of this compound would be the precursor of the pyridine N and that the primary alcohol, once activated, would allow incorporation of the benzimidazole moiety. Starting from 2-methyl-1-penten-3-one **14**, addition – elimination with pyrrolidine in the presence of acid afforded **15**. This material was treated first with oxalyl chloride to effect the cyclisation and then with methanol to convert the acid chloride to the methyl ester **16**. Reduction of the ester to the alcohol **13** set up the pyrone ready for conversion to the pyridone system by aminolysis with ammonia. Aromatisation by treatment of **17** with POCl$_3$ gave the pyridine **18** which reacted with 5-methoxy-mercaptobenzimidaole to yield the coupled product **19** from which omeprazole was accessible in two steps – displacement with methoxide (generated from methanol and KOH in DMSO) followed by S-oxidation as above.

5.4 Lansoprazole

Lansoproazole **20**, a close analogue of omeprazole, has found clinical success without any clear cut benefit over the latter. Its synthesis follows closely that of **1** but these details are much clearer (Scheme 5.4). Nucleophilic substitution by trifluoroethoxide on the nitropyridine-N-oxide **21** gave **22** which was then activated at the 2-position by acetylation of the N-oxide followed by rearrangement. Elaboration of this intermediate proceeded uneventfully to yield lansoprazole as a racemic mixture.

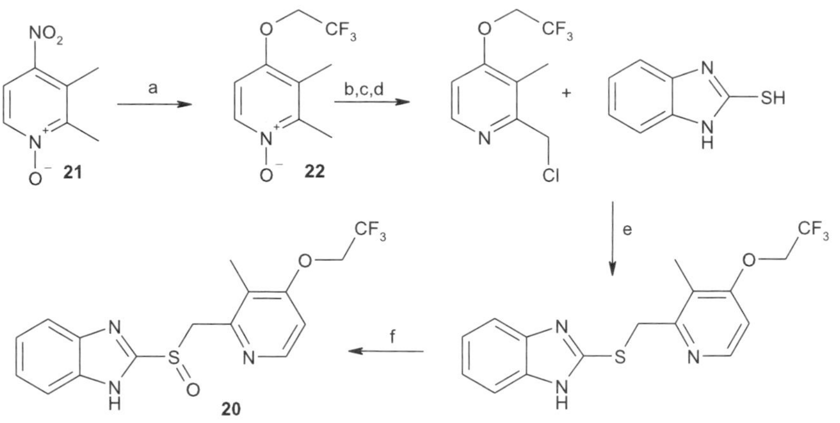

Scheme 5.5.   Reagents: (a) KOBu^t, rt, CF$_3$CH$_2$OH then heat 85°, 22 h; (b) H$_2$SO$_4$ (cat.), Ac$_2$O, 110°, 4 h; (c) MeOH, NaOH, rt, 30 min; (d) CHCl$_3$, SOCl$_2$, reflux, 30 min; (e) MeOH, MeO$^-$, 2-mercaptobenzimidazole, reflux, 30 min; (f) CHCl$_3$, mcpba, 5°, 30 min.

References:

Omeprazole:
U. F. Junggren and S. E. Sjostrand, European Patent, 1981, 0,005,129
P. Lindberg et al., *Medicin. Res. Rev.*, 1990, **10**, 1

Lansoprazole:
A. Nohara and Y. Maki, US Patent, 1986, 4,628,098
K. Kubo et al., *Chem. Pharm. Bull.*, 1990, **38**, 2853.

6 Modulation of central serotonin in the treatment of depression

6.1 Introduction

Major depression is an extremely prevalent disease affecting over 6% of the world's population and accounting for over $40 billion a year in direct health care costs and lost productivity in the US alone. Quite by chance, as an observation in patients taking iproniazid for tuberculosis, it was shown that inhibitors of monoamine oxidase (MAO, iproniazid was subsequently shown to have this activity) had clinical antidepressant activity by raising levels of central monoamines such as noradrenaline (NA), dopamine (DA) and serotonin (5-hydroxytrypatmine, 5-HT). Thus, antidepressant efficacy may be achieved by several different pharmacological activities including inhibition of noradrenaline uptake, serotonin uptake and monoamino oxidases. As an example, the postsynaptic activity of 5-HT released after stimulation of serotonergic presynaptic neurones is limited by reuptake into the presynaptic structures; blockade of this process will act to raise and/or prolong the activation of the postsynaptic receptors and hence serotoninergic neurotransmission. Fluoxetine **1** is one of several drugs which selectively potentiate the central actions of serotonin in this way, and unlike earlier antidepressants such as the tricyclics imipramine and clomipramine, exerts little direct activity at other receptors resulting in a lower occurrence of sedative and cardiovascular side effects.

6.2 Fluoxetine

A Mannich reaction on acetophenone **2** (Scheme 6.1) provided the key starting material **3** for the first reported synthesis of fluoxetine. Reduction of the ketone with diborane and chlorination of the resulting secondary alcohol **4** set up the reactive benzylic chloride for nucleophilic displacement with 4-trifluoromethylphenoxide. Von Braun degradation of the N,N-dimethylamine afforded the desired drug via the intermediacy of the N-cyano compound **5**.

Whilst fluoxetine **1** is marketed as the racemate, it was important to prepare and study the pharmacological behavior of the separated enantiomers. One route (Scheme 6.2) involved the asymmetric reduction of 3-chloropropyl aryl ketones with diisopinocampheylchloroborane to afford the (S)-(–)-alcohol **6** which was reacted first with NaI and then with aqueous methylamine. The resulting amine **7** was treated with NaH to generate the alkoxide and this nucleophile smoothly displaced fluoro- from 4-fluorobenzotrifluoride. Recrystallisation of the hydrochloride salt afforded (S)-(–)-fluoxetine the

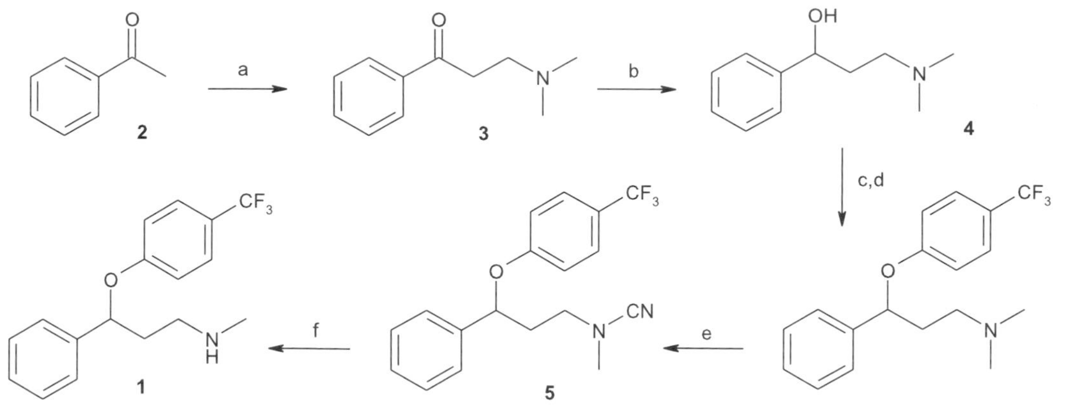

Scheme 6.1. Reagents: (a) $H_2C{=}O$, $NHMe_2$; (b) B_2H_6, THF, rt, 16 h; (c) HCl gas in $CHCl_3$, $SOCl_2$, reflux 5 h; (d) p-CF_3-C_6H_4-OH, NaOH, MeOH, reflux 5 days; (e) CNBr, PhH, PhMe, 5° then rt 16 h; (f) KOH, water, ethylene glycol, 130°, 20 h.

stereochemical integrity of which was confirmed by conversion to the diastereomeric urea **8** with (R)-1-(1-naphthyl)ethyl isocyanate. HPLC analysis of **8** showed two peaks in the ratio 96:4 indicating that racemisation had not occurred during synthesis. (R)-fluoxetine was prepared by classical resolution of fluoxetine using D-(−)-mandelic acid. Surprisingly, it was shown that the two isomers displayed little difference *in vitro* (K_i as inhibitors of the serotonin uptake carrier in rat cortex were 21 and 33 nM for the (+)- and (−)-isomers respectively) and in behavioral tests *in vivo*. Nevertheless, it is (S)-fluoxetine that is the predominant therapeutic enantiomer since this isomer is eliminated more slowly than the (R)-isomer in human beings.

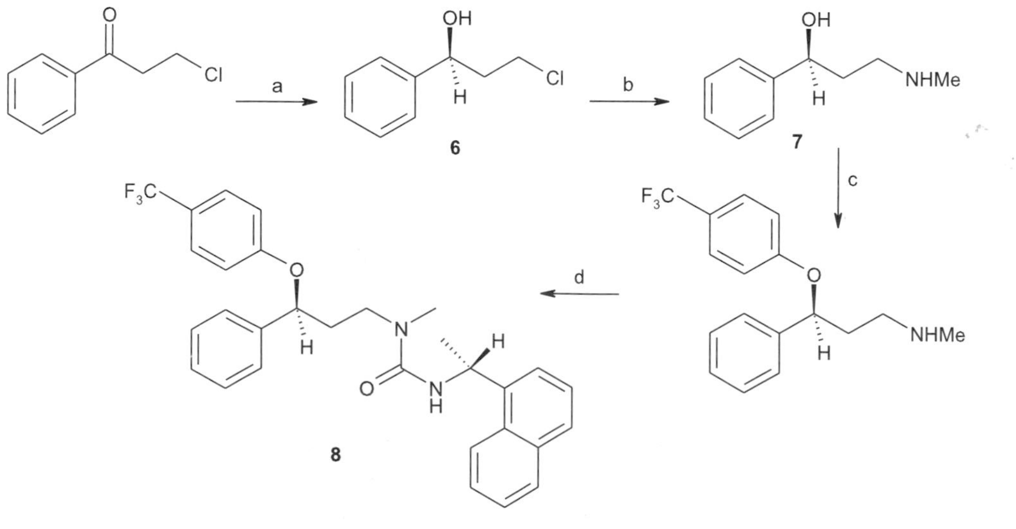

Scheme 6.2. Reagents: (a) diisopinocamphenylchloroborane; (b) NaI, Me_2CO, 16 h reflux then $MeNH_2$, H_2O, THF, rt, 16 h; (c) NaH, DMA, 70°, 0.5 h then p-trifluoromethylbenzofluoride, 90°, 2.25 h; (d) (R)-(−)-1-(1-naphthyl)ethyl isocyanate, PhMe, reflux, 2 h.

Scheme 6.3. Reagents: (a) L-(+)-diisopropyl tartrate, ; (b) D-(–)-diisopropyl tartrate; (c) DME, Red-Al, PhMe, 0–20° over 3 h; (d) Et₃N, MsCl, Et₂O, –10°, 3 h; (e) MeNH₂, H₂O, THF, 65°, 3 h; (f) DMA, NaH, 90°, 1.5 h then 4-chlorobenzotrifluoride, 100°, 2.5 h.

An alternative procedure which gave access to both enantiomers relied on the asymmetric epoxidation of cinnamyl alcohol and subsequent regioselective reduction of the resulting epoxide with sodium bis(2-methoxyethoxy)aluminium hydride ('Red-Al', Scheme 6.3). Either chiral epoxide **9a** or **9b** is directly available from cinnamyl alcohol by correct choice of epoxidation catalyst. (S)-(–)-fluoxetine required the use of D-(–)-diisopropyl tartrate to yield **9a** which was reduced to afford the 1,3-diol **10** in a 14:1 mixture with the 1,2-diol after much experimentation. The crude diol was reacted selectively at the primary alcohol with 1 equivalent of mesyl chloride, the product purified by chromatography and the mesyl group displaced with methylamine in aqueous THF. Generation of the alkoxide of the secondary alcohol and reaction with p-chlorobenzotrifluoride completed the synthesis. The (R)-(+)-isomer was prepared in a similar manner but starting with **9b**.

Recent debate has focussed on the (R)-isomer as being the preferred drug and, indeed, this single enantiomer is currently being developed based on the argument that the prolonged duration of the (S)-isomer in man can contribute to the occasional side effects noted since the concentration of that isomer in the body may well build up after multiple doses. As an interesting foot note to the patent that provokes this debate, the claim is to R-(–)-fluoxetine but it should be noted that this material, as the HCl salt, is laevorotatory in methanol but dextrorotatory in water.

6.3 Sertraline

Early research showed that *trans*-N-alkyl-4-phenyl-tetrahydronaphthylamines e.g. **11** had excellent antidepressant activity acting primarily by blocking synaptosomal noradrenaline uptake – and thus associated with the typical range of mechanism-based side effects. The *cis*-racemates were mainly inactive as noradrenaline uptake blockers and this discouraged detailed exploration within the *cis*-series. As the serotonergic hypothesis became established, these and other analogues were evaluated for the now preferred mechanism and it took almost 8 years to discover that the *cis*-series, typified by sertraline, *cis*-(+)- **12**, had the better profile. Unlike fluoxetine, sertraline is a single enantiomer with *cis*-(1S,4S) absolute configuration.

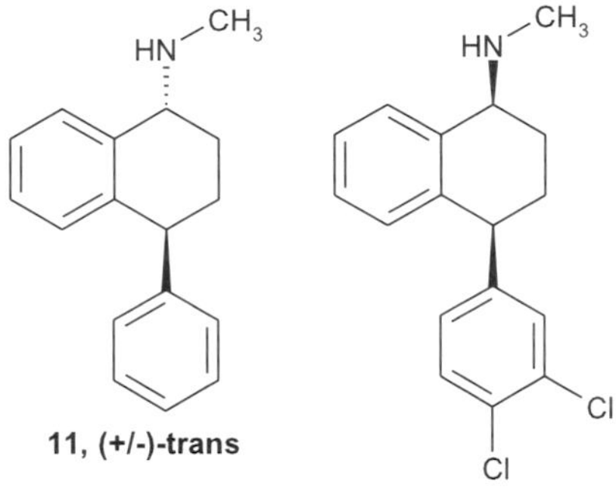

11, (+/-)-trans

(+)-cis-12

The starting benzophenone **13** was readily available on a large scale by AlCl$_3$ mediated Friedel-Crafts acylation of benzene with 3,4-dichlorobenzoyl chloride (Scheme 6.4). A Stobbe reaction between benzophenone **13** and diethylsuccinate afforded the mono acid **14** which was hydrolysed and decarboxylated under strongly acidic conditions and then hydrogenated to the

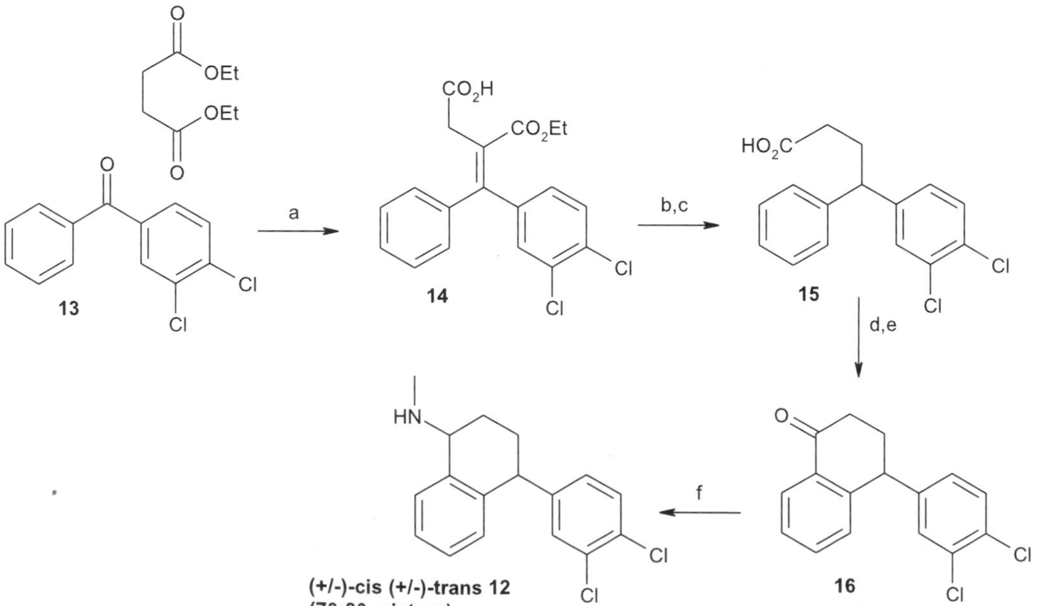

Scheme 6.4. Reagents: (a) t-BuOH, t-BuO$^-$, reflux, 16 h; (b) HBr, HOAc, reflux, 36 h; (c) EtOAc, H$_2$, 5% Pd-C; (d) SOCl$_2$, PhMe, reflux, 1.25 h; (e) AlCl$_3$, CS$_2$, 16 h, rt; (f) MeNH$_2$, TiCl$_4$, 10° to rt then H$_2$, 10% Pd-C.

Scheme 6.5. Reagents: (a) AlCl$_3$, 60°, 2.5 h; (b) aq. NaOH, NaBH$_4$, 80°, 2.5 h; (c) PhH, conc. H$_2$SO$_4$, 95° 1h then 140°, 1.5 h.

4,4-diarylbutanoic acid **15**. Exposure to thionyl chloride afforded the acid chloride which was induced to cyclise under Friedel-Crafts conditions on to the more reactive aryl ring. In the case of analogues bearing electron donating groups in place of the 3,4-dichloro arrangement, an alternative process was required since the switch in reactivity gave the undesired cyclisation product. The substituted tetralone **16** was condensed with methylamine in the presence of the Lewis acid, titanium tetrachloride, and the resulting imine immediately reduced to a mixture of *cis-* and *trans*-amines **12**. Using 10% Pd-C as the hydrogenation catalyst gave a 70:30 mixture in favour of the *cis*-form and this was purified as its hydrochloride salt by fractional crystallisation. The pure *cis*-racemate was resolved with D-(–)-mandelic acid to give the (+)-(1S,4S) isomer of **12** whose absolute configuration was determined by X-ray crystallography. The alternate enantiomer, *cis*-(–)-(1R,4R), could be obtained using L-(+)-mandelic acid in place of the D-form.

The intermediate tetralone **16** could be made by a more efficient process involving a Friedel-Crafts reaction (Scheme 6.5) using three equivalents of AlCl$_3$ and 1,2-dichlorobenzene which was acylated with succinic anhydride to give **17** accompanied by only trace amounts of the ortho-regioisomer. Borohydride reduction of the ketone group and spontaneous lactonisation during the work up gave **18**. After much experimentation with a variety of protic and Lewis acids, it was shown that **18** could be converted directly into **16** by alkylation – acylation with benzene in the presence of concentrated sulphuric acid presumably through the intermediacy of the diaryl acid **19**.

More recently, an enantioselective synthesis of sertraline has been described based on the regio- and enantioselective ring opening of oxabenzonorbornenes (such as **20**, Scheme 6.6) under reductive conditions. Thus, **20**, readily available from a Diels-Alder reaction between benzyne and furan, was treated with DIBAL in the presence of Ni(COD)$_2$ and (S)-BINAP, preferably in toluene, to afford **21** in good yield and 91% ee recrystallisation of which improved the ee to 98%. Protection of the alcohol followed by bromination of the double bond and elimination of HBr with diazbicycloundecane (DBU) gave **22**. A Stille cross coupling reaction with 3,4-

Scheme 6.6.   Reagents: (a) n-BuLi, Et$_2$O, −78°; (b) Ni(COD)$_2$, (S)-BINAP, DIBAL-H, −40°; (c) TBDPSCl, imidazole, DCM, DMAP, rt; (d) Br$_2$, DCM, 0°; (e) DBU, PhH; (f) (MeCN)$_2$PdCl$_2$; AsPh$_3$, (3,4-diCl)C$_6$H$_3$SnMe$_3$, NMP, 80°, 1.5 h; (g) TBAF, THF, AcOH, 3 days; (h) [Ir(COD)pyPCy$_3$]PF$_6$, H$_2$, 1000 psi, DCM; (i) DPPA, DBU, THF; (j) H$_2$, Pd-C, EtOH; (k) ClCO$_2$Et, MeCN, K$_2$CO$_3$, then LiAlH(OMe)$_3$, THF, reflux, 40 h.

dichlorophenyltrimethylstannane and deprotection gave **23** with the caveat that acetic acid was present to avoid aromatisation following elimination of silanol. The next step required hydrogenation of the double bond directed by the secondary alcohol and this was achieved by hydrogenation catalysed with an iridium complex know as 'Crabtree's' catalyst with high facial selectivity to give **24** in a ratio of 28:1 trans:cis. Complete inversion of stereochemistry during formation of the azide **25** gave, after catalytic reduction, the enatiomerically pure primary amine but now with the ring substituents having a *cis* relationship. This was treated with ethyl chloroformate and the resulting urethane reduced directly to sertraline.

Another approach involved catalytic asymmetric synthesis to produce the tetralone **16** with 100% ee of the (S)-enantiomer which could be transformed into (+)-cis-sertraline by reductive amination described above. Addition of the carbene generated from the diazo ester **26** (Scheme 6.7) in the presence of the Rh catalyst **27** to styrene afforded the cyclopropane **28** (94% ee) which was oxidatively degraded to the diester **29**. Homoconjugate addition of the cuprate shown, prepared *in situ* from 3,4-dichlorophenyl iodide, t-butyllithium and cuprous cyanide, gave the addition product **30** which was hydrolysed and decarboxylated to the corresponding diarylbutyric acid. This intermediate was cyclised with chlorosulphonic acid to the tetralone of 100% ee.

6.4   Paroxetine

Paroxetine **31** is the third selective serotonin uptake inhibitor that we shall consider. The early chemistry is characterised by classical methods of obtaining a single enantiomer in a structure that contains four diastereomers –

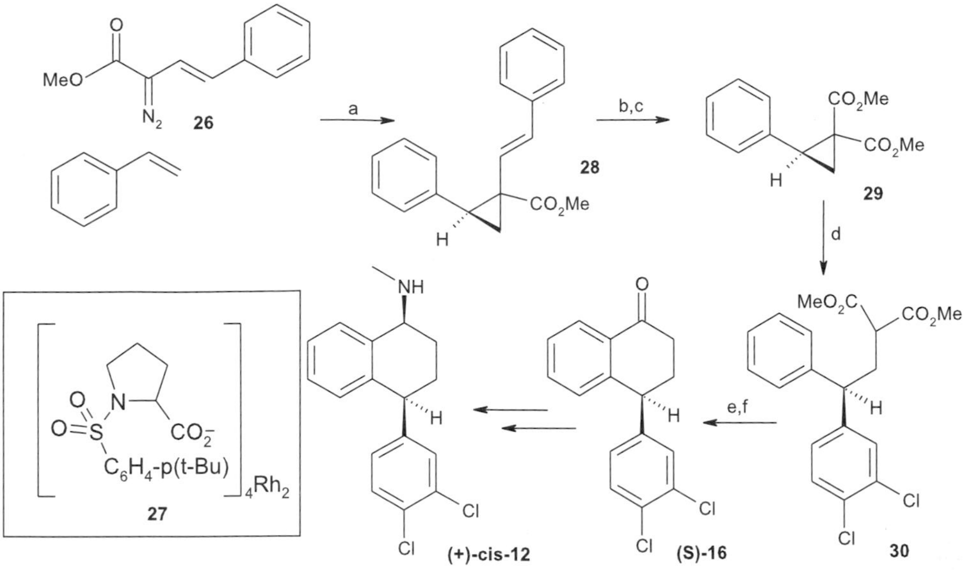

Scheme 6.7. Reagents: (a) Rh catalyst 27, pentane, 0°, 2 h, recryst. 94% ee; (b) KMnO₄, NaIO₄, t-BuOH, H₂O, 0.5 h, rt; (c) DMS, K₂CO₃, Me₂CO, 3 h, rt; (d) (3,4-diClC₆H₄)₂CuLi₂CN, rt, 1 h; (e) 6N HCl, reflux, 20 h; (f) ClSO₃H, DCM, 0.5 h, rt.

procedures which are unattractive on a manufacturing scale given the wastage of material concerned. Of the four isomers, it is the (–)-*trans*-(4R,3S)-form which is the active drug:

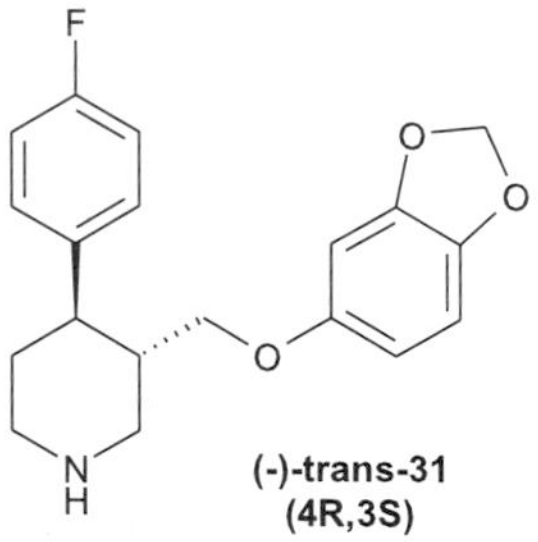

The first synthesis of paroxetine and other analogues within the series used a route wherein an intermediate allowed ready separation of the *cis*- and *trans*-forms each as the racemate (Scheme 6.8). Starting with arecoline **32**, Michael addition of 4-fluorophenyl magnesium bromide gave **33** as a mixture of *cis*-(**33a**) and *trans*-(**33b**). It was found that the *cis*-form could be isolated pure by equilibration of the mixture with NaOMe in refluxing benzene. The *trans*-form was available only by separation of the original mixture. The cis material was hydrolysed with aqueous HCl and the resulting acid treated with thionyl chloride to give **34**. Transformation into diastereomeric menthol esters **35** allowed the isomers to be separated by distillation. The (–)-*cis*-form was hydrolysed under acidic conditions and reduced to the carbinol **36** which was

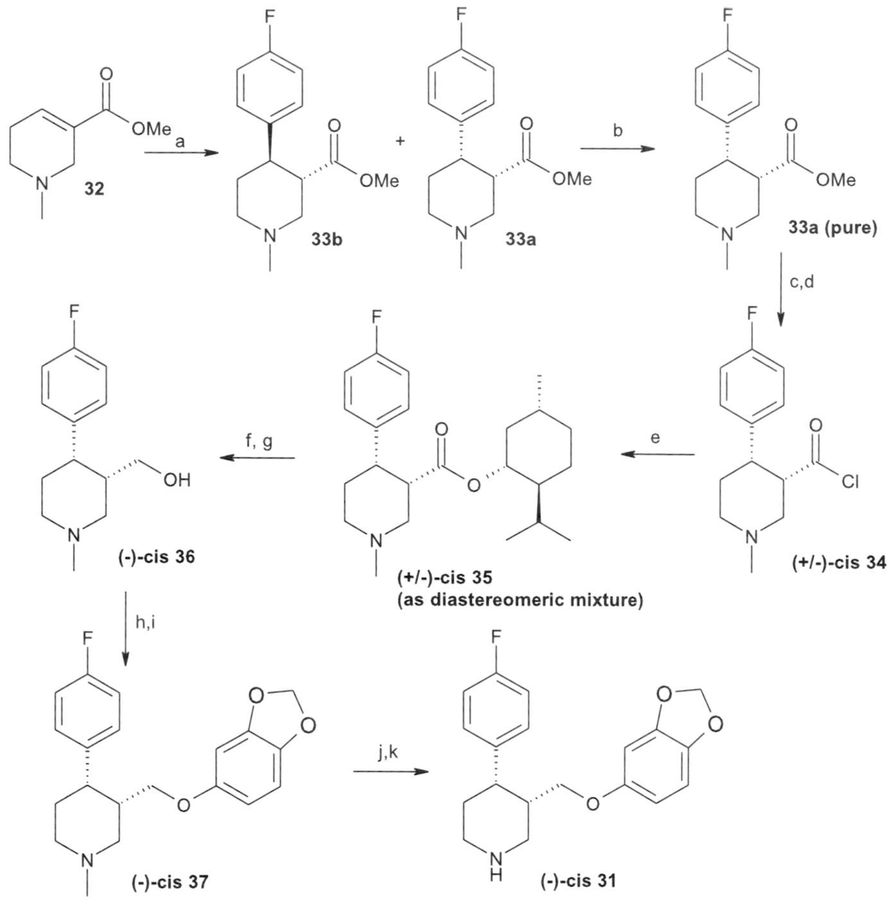

Scheme 6.8. Reagents: (a) 4-F-C$_6$H$_4$MgBr; (b) NaOMe, PhH, 2 h, reflux; (c) aq. HCl; (d) SOCl$_2$, 3 h, rt; (e) (−)-menthol, pyridine, 0°, then 16 h rt; (f) fractional distillation to give the single diastereomer; (g) LAH; (h) SOCl$_2$, CHCl$_3$, 10° then reflux 6 h; (i) Na, MeOH, 3,4-methylenedioxyphenol, reflux 16 h; (j) DCM, PhOCOCl, 0° then 16 h at rt; (k) KOH, methylcellusolve, reflux, 4 h.

chlorinated with thionyl chloride and etherified with 3,4-methylenedioxyphenol as its sodium salt. Finally, it was necessary to remove the N-methyl group of **37** and this was achieved by treatment first with phenylchloroformate and then hydrolysis and concomitant decarboxylation of the urethane under forcing conditions. Subsequently, it was found that the use of vinylchloroformate allowed the demethylation process to occur under much milder conditions. Thus the intermediate vinylurethane (**38**, Scheme 6.9) could be hydrochlorinated with HCl gas and the chloroethylurethane **39** hydrolysed in refluxing methanol.

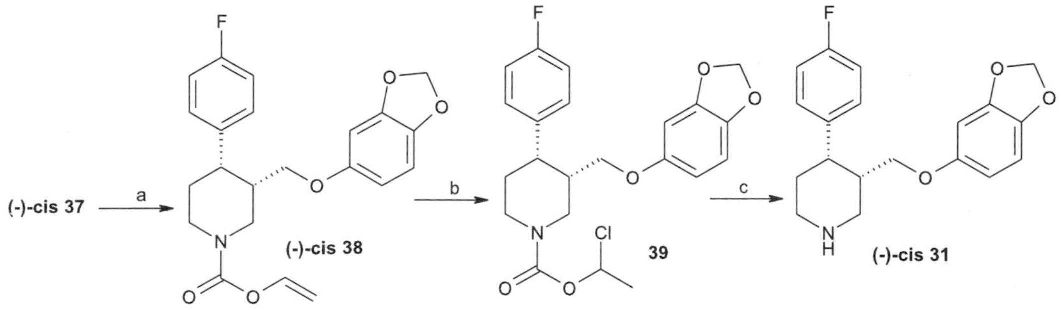

Scheme 6.9. Reagents: (a) DCM, CH$_2$=CHOCOCl, 0°, then rt 3 h; (b) dry HCl gas, DCM; (c) MeOH, 1 h, reflux.

References:

Fluoxetine
B. B. Molloy and K. K. Schmiegel, US Patent 1982, 4,314,081
D. W. Roberston et al., *J. Medicin. Chem.* (1988) 31, 1412–1417
Y Gao and K. B. Sharpless, *J. Org. Chem.*, 1988, **53**, 4081

Sertraline
W. M. Welch et al., US Patent 1985, 4,536,518
M. Williams and G. Quallich, *Chem. and Ind. (Lond.),* 1990, 315
M. Lautens and T. Rovis, *J. Org. Chem.*, 1997, **62**, 5246
E. J. Corey and T. G. Gant, *Tet. Lett.*, 1994, **35**, 5373.

Paroxetine
J. A. Christensen and R. F. Squires, US Patent, 1977, 4,007,196

7 Hypnotic, anxiolytic anticonvulsant and muscle relaxant agents acting at the benzodiazepine receptor

7.1 Introduction

The discovery of ligands acting at benzodiazepine receptors is one of many stories where a succession of serendipitous events led to the final breakthrough. The first was the decision to explore novel chemical structures away from the barbiturates and urethanes that represented the standard drugs used at the time, albeit acting by a mechanism then unknown. In a campaign to prepare new benzoheptoxadiazines (e.g. **1**, Scheme 7.1) based on earlier chemistry that was subsequently shown to be incorrect, chlordiazepoxide **2**

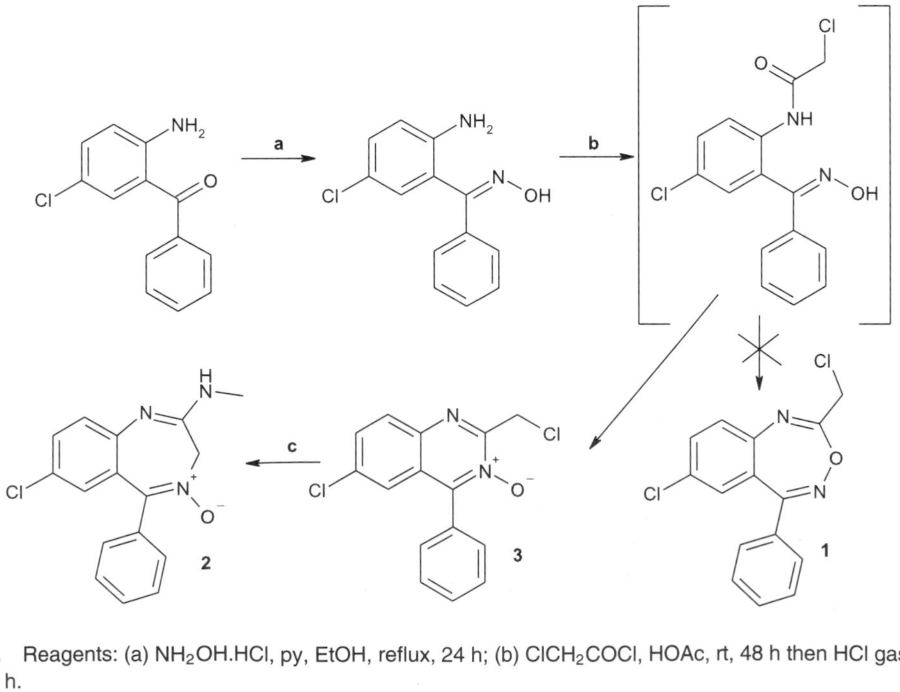

Scheme 7.1. Reagents: (a) NH$_2$OH.HCl, py, EtOH, reflux, 24 h; (b) ClCH$_2$COCl, HOAc, rt, 48 h then HCl gas; (c) 30% MeNH$_2$ in MeOH, rt, 15 h.

was produced which had remarkable activity in animal models predictive of anxiolytic behaviour and was introduced in 1960. The benzoheptoxadiazine structure was shown to be a quinazoline N-oxide **3** and this, when reacted with a primary amine, underwent nucleophilic attack at the C-2 center followed by ring expansion to afford the active drug **2**. By 1963, diazepam was also available followed, over the next decade, by well over 20 other benzodiazepines all finding a role based on their level of efficacy (see below), metabolic fate and duration of action.

It is now known that benzodiazepines exert their pharmacological effects by interaction at a specific site on the $GABA_A$ receptor complex. The $GABA_A$ receptor belongs to the superfamily of ligand gated ion channels – here the ligand is GABA and the ion channel is a chloride ion channel. The receptor is composed of a hetero-pentameric protein with each monomer (termed α, β γ and δ) proposed to have four helical domains which become embedded in the cell membrane and together form a pore, or channel. Given that there are multiple sub-types of each sub-unit, the number of variations in the final complex is immense. Detailed mutational analysis has indicated that the binding site for benzodiazepines (and newer, structurally unrelated ligands) is located on the alpha sub-unit and it is now possible to distinguish ligands that have selectivity for a specific alpha sub-unit. Thus, GABA released from a presynaptic neurone activates the postsynaptic GABA receptor and this mediates Cl^- channel opening which hyperpolarises the cell membrane and causes inhibition of neuronal activity. Depending on the level of efficacy displayed by a given benzodiazepine, the ligand may act as an agonist to enhance the effect of GABA, as a partial agonist to produce an intermediate effect, or as an antagonist which results in no effect other than blocking the benzodiazepine receptor and thus preventing other ligands from interacting at that site. Confused? To add to the complication, it is also possible to synthesise benzodiazepines that act as 'inverse agonists' which act to inhibit the GABA-mediated response and hence behave as anxiogenic agents.

For some time it was hoped that the level of efficacy in benzodiazepine ligands could be titrated to enable new anxiolytic agents to be produced devoid of the side effects common with this class of drugs such as drowsiness, ataxia and dependence. In fact, these are most readily controlled by selection of the drug with the most appropriate combination of half life and metabolism either to active or inactive metabolites. As an example, diazepam **4** (long half life) is demethylated in the liver to an active metabolite (very long half life, Scheme 7.2) and in turn this is oxidised to oxazepam **5** (also active but with short duration). For anxiolytic agents, a relatively long duration is preferred to avoid multiple daily dosing, whereas a short duration drug is indicated for hypnotic activity. Other successful variants include lorazepam **6**, temazepam **7** and alprazolam **8**. The receptor antagonist flumezanil **9** has been used to reverse the effects of benzodiazepine-induced general anaesthesia. Finally, the current market leader for insomnia with sales in excess of $0.5 b is zolpidem, **10**.

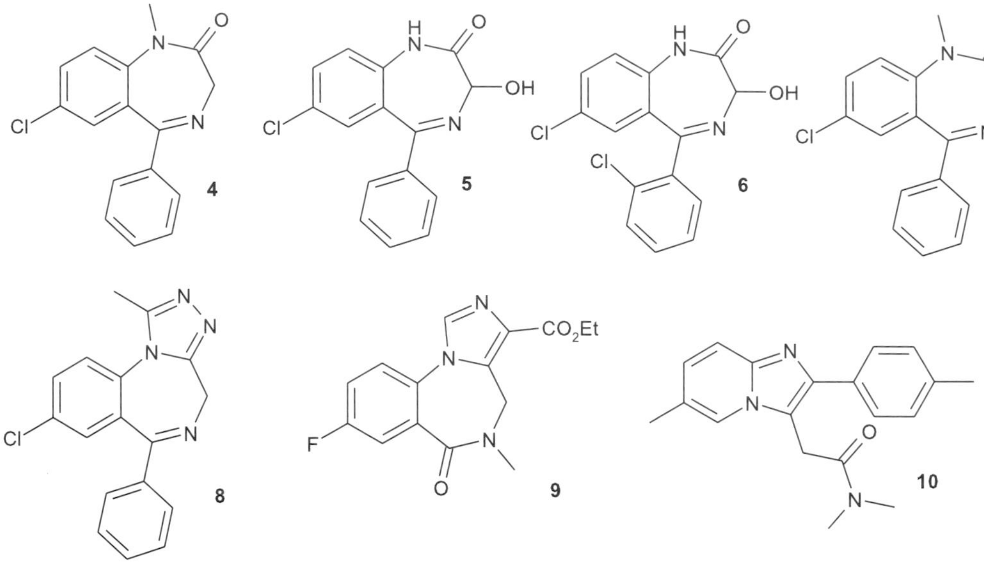

Scheme 7.2.

7.2   Diazepam

While chlordiazepoxide was in late stages of clinical development, it became necessary to study different formulations, particularly for pediatric and geriatric use, which would mask the bitter taste of the drug and would prevent decomposition of the drug once dissolved in water. The decomposition product **11** (Scheme 7.3) had the same level of activity as **2** and reduction of the N-oxide also afforded a highly active anxiolytic and anticonvulsant agent **12**. Thus the two features of chlordiazepoxide, the N-oxide and the basic entity, previously thought to be crucial for activity were now shown to be superfluous. Polonovsky rearrangement of the N-oxide in the presence of acetic anhydride followed by basic hydrolysis of the acetyl group gave the 3-hydroxy derivative **13** which was subsequently developed as oxazepam. As a final discovery from this rich area of chemistry, N-methylation of **12** gave diazepam which was significantly more active than chlordiazepoxide. The syntheses of lorazepam and temazepam follow similar routes – simple N-methylation of **13** with dimethylsulphate gives temazepam **7** directly.

Diazepam was also available by more convenient chemistry (Scheme 7.4) starting with the benzophenone **14** followed by acylation on the anilino-NH$_2$ with chloroacetyl chloride. Nucleophilic displacement of the chloro group in liquid ammonia gave the intermediate aminoacylamide **15** which could be cyclised upon heating. Alternatively, by using NH$_3$ in ethanol under reflux, cyclisation took place *in situ*.

A newer approach employed the isatoic anhydride **16** and glycine as the reagent inputs (Scheme 7.5). Thus **16** was first alkyated with methyl iodide and the product treated with the anion of glycine generated by addition of

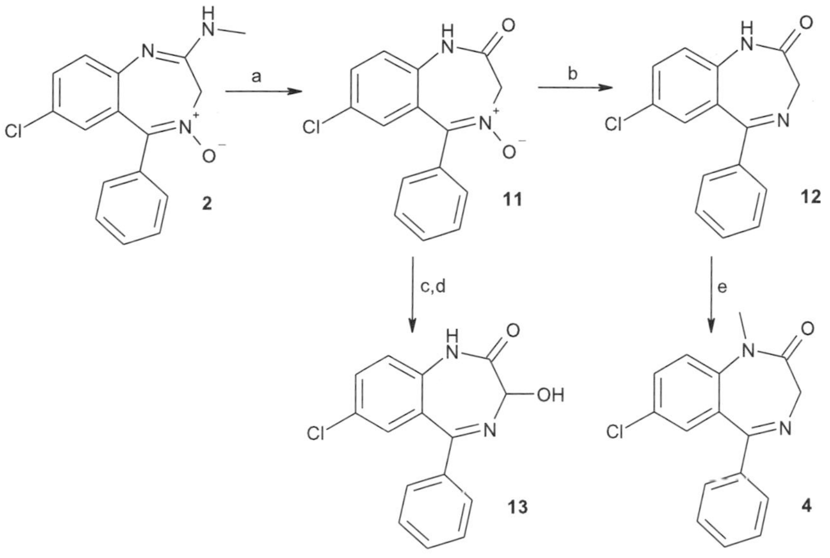

Scheme 7.3. Reagents: (a) HCl salt, H$_2$O, 30 days, rt; (b) PCl$_3$, CHCl$_3$, reflux, 30 min; (c) Ac$_2$O, steam bath, 20 min.; (d) EtOH, 4N NaOH, rt; (e) NaOMe, PhH, reflux under Dean & Stark conditions then Me$_2$SO$_4$, reflux, 1 h.

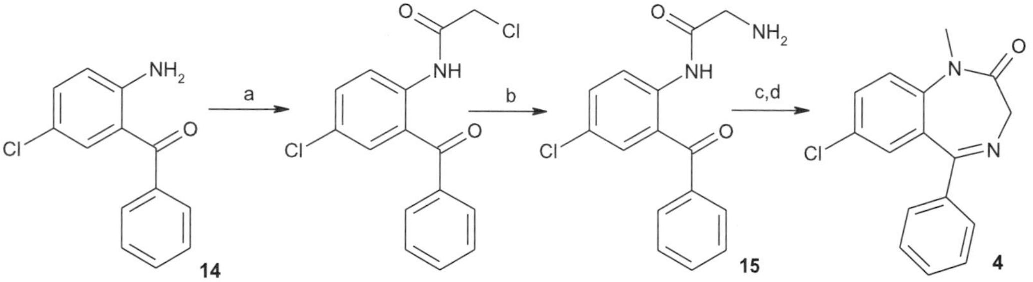

Scheme 7.4. Reagents: (a) Dioxane, ClCH$_2$COCl, 10°, 1 equiv. NaOH, 30 min; (b) liq. NH$_3$, reflux 5 h; (c) pyridine, reflux 1 h; (d) MeOH, NaOMe, evaporate then DMF, MeI, rt, 30 min.

exactly one equivalent of triethylamine to afford the benzodiazepinedione **17** directly. In order to avoid the poor yields observed with Grignard addition to **17**, it was first necessary to form the acetyl derivative **18** and this reacted smoothly with PhMgBr to give **19**. Oxime formation and ring closure with sodium hydrogen sulphite gave diazepam possibly via the intermediacy of the sulphamate (C = N-SO$_3$Na).

Palladium catalysed carbonylation of aryl halides formed the basis of a general procedure towards 1,4-benzodiazepines including diazepam (Scheme 7.6). p-Chloroaniline was selectively brominated *ortho* to the NH$_2$ group which was then methylated and acylated with Z-glycine acid chloride to give **20**. For the carbonylation reaction, it was first necessary to change N-protecting groups from benzyloxycarbonyl to benzyl. The resulting secondary

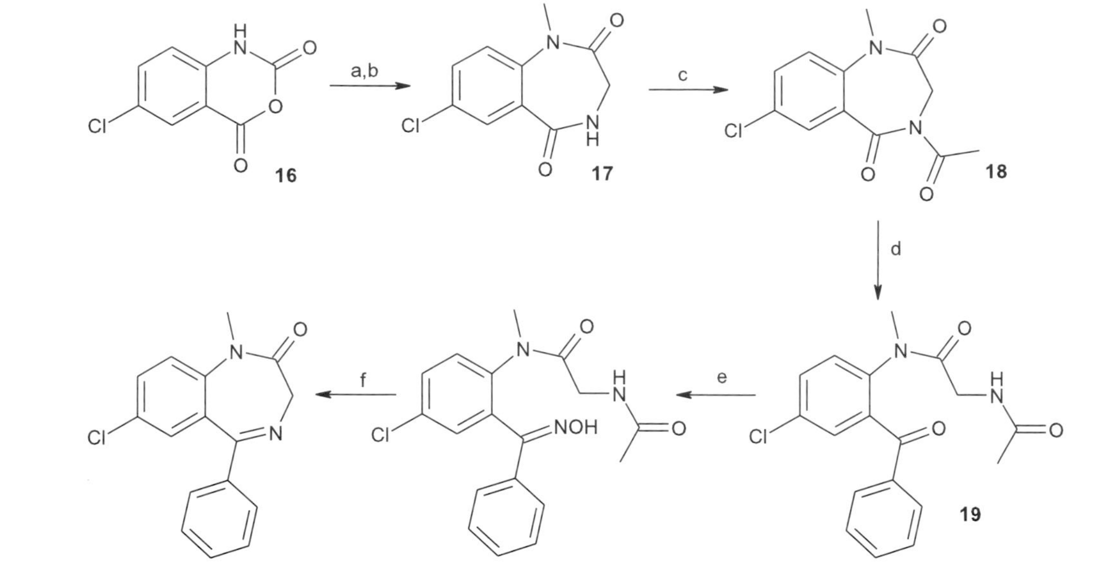

Scheme 7.5. Reagents: (a) MeI, DMF; (b) glycine, Et₃N, H₂O, rt, 5 h; (c) Ac₂O, reflux, 2.5 h; (d) dry THF, PhMgBr, rt, 1.5 h; (e) NH₂OH.HCl, Py, 70°, 45 h; (f) NaHSO₃, EtOH, H₂O, reflux, 12 h.

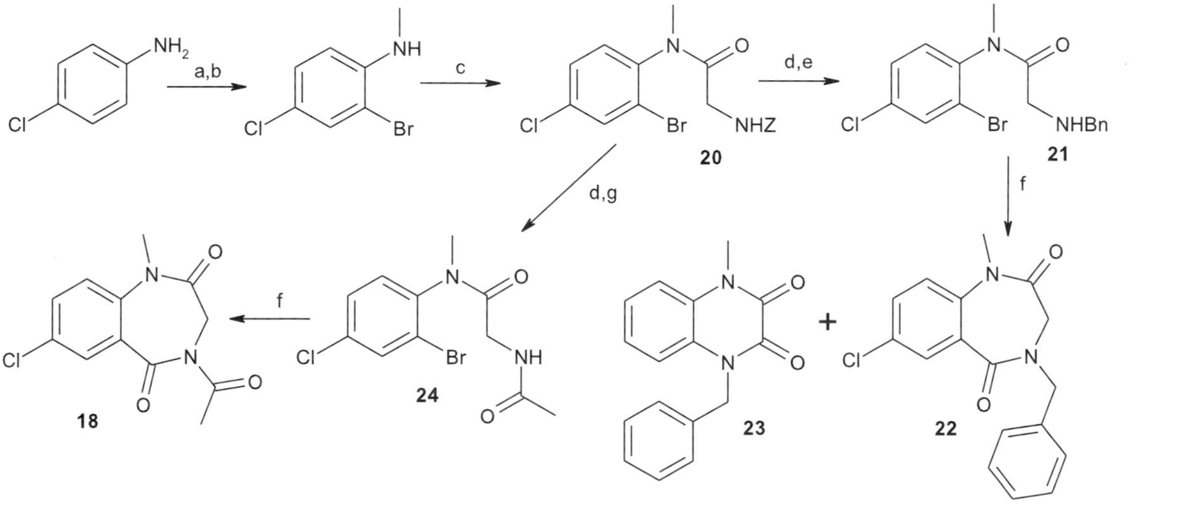

Scheme 7.6. Reagents: (a) NaOAc, HOAc, Br₂, 1 h, rt; (b) MeI, DCM; (c) Z-glycine, Et₂O, PCl₅, rt, 1h then add aniline; (d) DCM, HBr, HOAc, rt, 1 h; (e) PhCHO, PhH, rt, 4 h then NaBH₄, MeOH, rt, 1 h; (f) Ph₃P, nBu₃P, Pd(OAc)₂, HMPA, CO, 5 atm., 100°, 40 h; (g) K₂CO₃, DCM, AcCOCl, 0°, 30 min then 1 h, rt.

amine **21**, obtained by reductive alkylation of the primary amine with benzaldehyde, was reacted with carbon monoxide in the presence of palladium acetate and Ph₃P to give the diazepinedione **22** with **23** as a biproduct which was also formed in the absence of catalyst. The latter was the result of substitution of the bromine atom directly by the amino group followed by air oxidation and was the major product under most experimental conditions. However, the N-acetyl intermediate **24** smoothly generated **18** with no sign of the corresponding biproduct. and **18** and **22** represent convenient relays toward diazepam as described above.

7.3 Alprazolam

The benzodiazepinone **12** and thione **25** are versatile intermediates en route to a wide range of tricyclic systems of which alprazolam **8** is but one example (Scheme 7.7). Thus **25** was reacted with acetylhydrazide in refluxing methanol and then heated at 200° to give **8**. In an alternative approach to avoid the vigorous reaction conditions and to improve the yield of the cyclisation reaction, the hydrazone **26** was studied. The hydrazine **27** was treated with acetaldehyde and the resulting hydrazone heated in benzene in the presence of diethyl azodicarboxylate to give **8**.

Yet another method, formally involving N-insertion into the triazoloquinoline **28**, started with the 2-chloroquinoline **29** (Scheme 7.8). The chloro group was readily displaced by hydrazine and the hydrazide cyclised with triethylorthoacetate to afford **28**. Oxidation of this intermediate, preferentially

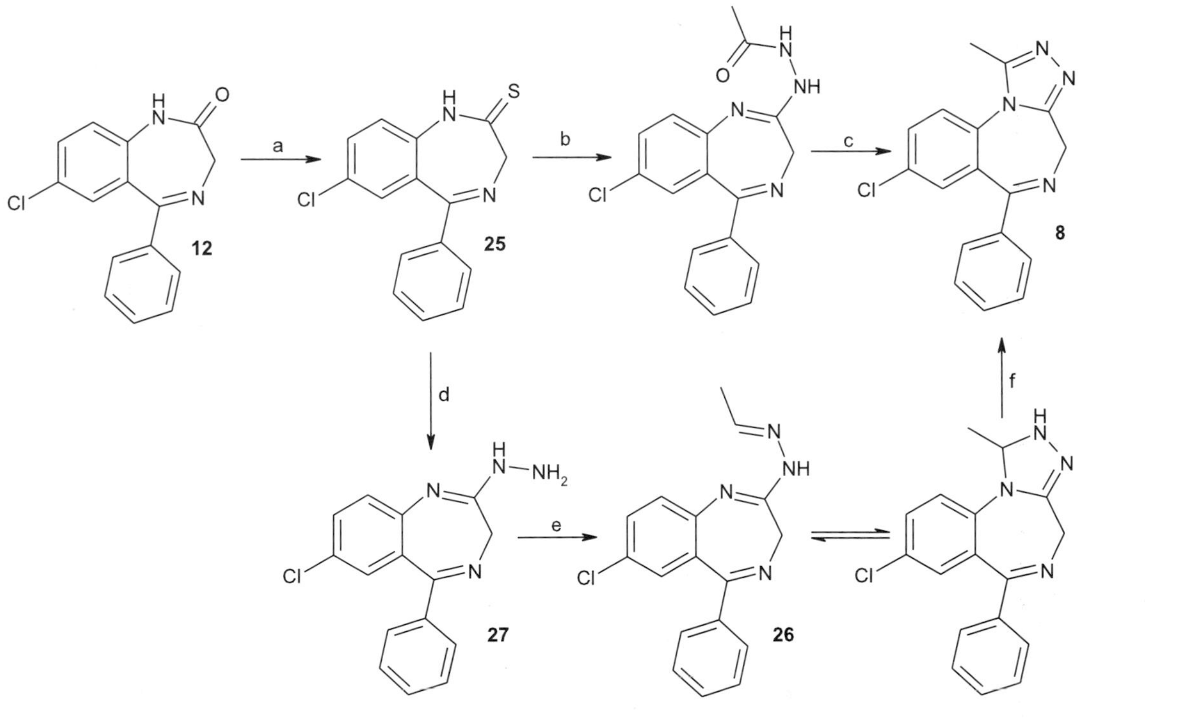

Scheme 7.7.   Reagents: (a) Py, P_2S_5, reflux, 30 min; (b) acetylhydrazide, MeOH, reflux 24 h; (c) 200° under reduced pressure (12 mm) until bubbling ceases; (d) $NH_2NH_2.H_2O$, MeOH, reflux; (e) CH_3CHO, THF, AcOH, warm, 5 min; (f) PhH, $EtO_2CN = NCO_2Et$, reflux 16 h.

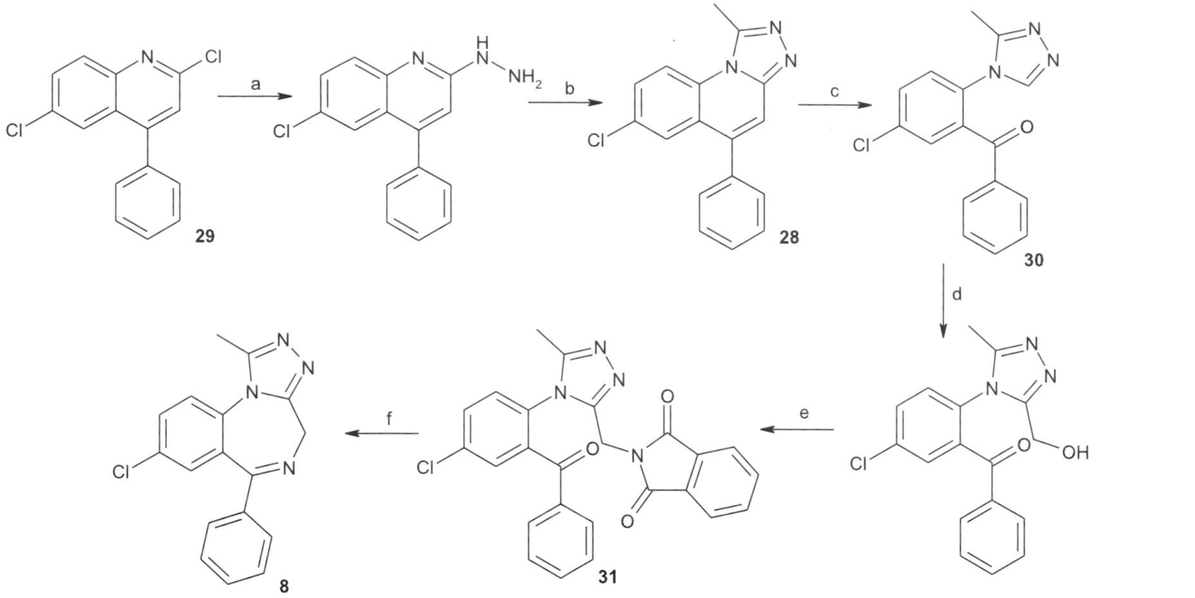

Scheme 7.8.   Reagents: (a) $NH_2NH_2.H_2O$, reflux, 1 h; (b) $CH_3C(OEt)_3$, xylene, reflux, 3 h; (c) Me_2CO, K_2CO_3, H_2O, $NaIO_4$, $KMnO_4$, 3 days; (d) $(CH_2 = O)_n$, xylene, reflux; (e) phthalimide, Ph_3P, EtO_2C-N = N-CO_2Et, THF, rt, 24 h; (f) EtOH, $NH_2NH_2H_2O$, rt, 2.5 h.

with permanganate – periodate gave **30** possibly via conversion of the expected aldehyde to the acid followed by spontaneous decarboxylation. The vacant triazole 2-position could be hydroxymethylated with formaldehyde at high temperature and this converted into the phthalimide **31** by a Mitsunobu reaction. Exposure of this intermediate to hydrazine gave alprazolam **8** directly.

7.5   Flumezanil

Earlier work with diazepam using isatoic anhydride as the starting material (Scheme 7.5) provided the route development for flumezanil **9** (Scheme 7.9), although, in this case, the conditions for the reaction with N-methylglycine differed significantly. Under these conditions, the intermediate **32** cyclised to the secondary lactam **33** and this was activated as the phosphate diester **34** and reacted with the anion derived *in situ* from ethyl isocyanoacetate to give **9** directly.

7.6   Zolpidem

Zolpidem **10** is a non-benzodiazepine hypnotic which binds at the same site of the α subunit of the GABA receptor complex as classical benzodiazepines. However, in contrast to the benzodiazepines, it binds selectively to one of three subtypes of benzodiazepine receptors, the 'ω_1 receptor', and this limits its loci of action within the brain relative to benzodiazepines. Although its

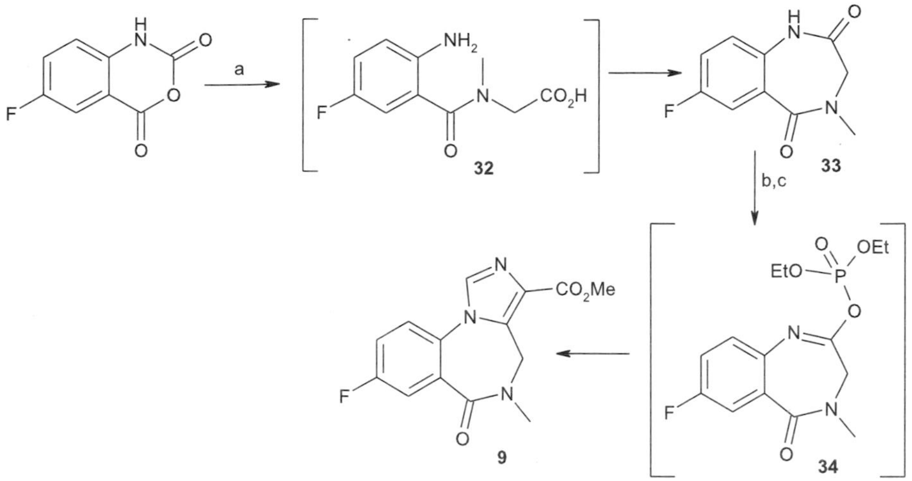

Scheme 7.9. Reagents: (a) DMSO, sarcosine.HCl, 100°, 1.5 h; (b) DMF, KOBu^t, 35°, 10 min. then −20°, ClP = O(OEt)$_2$; (c) (separately) DMF, KOBu^t, 40°, EtO$_2$CCH$_2$NC, 10 min then cool to −10° and add **34** to reaction mixture, 1 h, −20°.

selectivity is not absolute, this may explain why the drug is devoid of some side effects seen with other hypnotic benzodiazepines.

The synthesis started with 2-amino-5-methylpyridine which was reacted with the α-chloroketone **35** (Scheme 7.10) to generate the pyrrolo[1,2-a]pyridine **36** and this undergoes a Mannich reaction to introduce the functional sidechain ready for chain extension. Quaternisation of **37** with methyl iodide followed by nucleophilic displacement with cyanide gave **38**

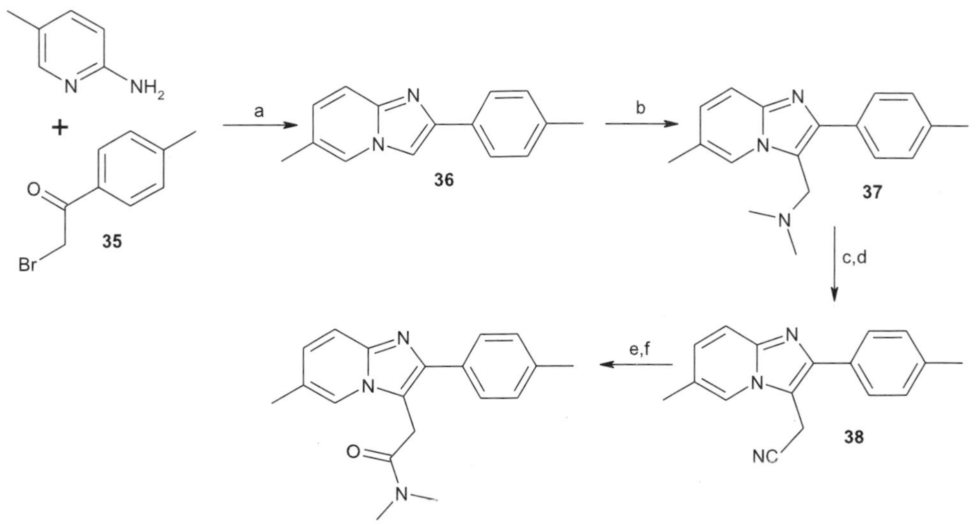

Scheme 7.10. Reagents: (a) EtOH, 3 h, 60°; (b) HOAc, Me$_2$NH, 40% CH$_2$ = O in H$_2$O, 60° 3 h; (c) MeI; (d) NaCN, H$_2$O; (e) 99% formic acid, dry HCl gas, 4 h, then KOH, EtOH, reflux, 10 h; (f) carbonyldiimidazole, THF, 20–40° then cool to 0°, Me$_2$NH, THF.

which was hydrolysed first to the primary amide under acidic conditions and then to the corresponding acid using KOH. This acid was activated with carbonyldiimidazole and coupled with dimethylamine to afford zolpidem.

References:

Chlordiazepoxide
L. H. Sternbach, US Patent 1959, 2,893,992
L. H. Sternbach and E. Reeder, *J. Org. Chem.*, 1961, **26**, 1111

Diazepam
E. Reeder and L. H. Sternbach, US Patent, 1968, 3,371,085
L. H. Sternbach, *J. Medicin Chem.*, 1979, **22**, 1
M. Gates, *J. Org. Chem.*, 1980, **45**, 1675

Alprazolam
J. B. Hester, US Patent, 1976, 3,987,052 (1976)
J. B. Hester, D. J. Duchamp and C. G. Chidester, *Tet. Lett.* 1971, 1609
J. B. Hester, *J. Het. Chem.*, 1980, **18**, 575

Flumezanil
M. Gerecke et al., US Patent 1982, 4,346,031

Zolpidem
J. P. Kaplan and P. George, US Patent, 1983, 4,382,938

8 Another histamine receptor: blockers of the histamine-1 receptor for the treatment of seasonal allergic rhinitis

8.1 Introduction

Histamine is released in a variety of allergic conditions such as seasonal rhinitis ('hay fever'), urticaria (rash), pruritis and insect stings as well as a reaction to certain drugs such as penicillin and aspirin. The reaction is the result of an inappropriate antibody response: IgE binds to the surface of mast cells which are widely distributed throughout the body after the first exposure to the antigen. After renewed exposure, degranulation of the mast cells causes release of a variety of pro-inflammatory mediators with histamine being the one of the first 'on the scene'. If the release is localised, then the symptoms associated with allergy become apparent; more general mediator release may cause anaphylactic shock which requires emergency treatment. It turns out that histamine mediates its activity by interaction at H_1 receptors which are GPCR's closely related to, but clearly distinct from, the H_2 receptors at which anti-ulcer drugs bind (see Chapter 2). Agents which selectively block these receptors are therefore of benefit in histamine-related allergic responses.

First generation H_1 blockers, which appeared in clinical practice during the 1940's, suffered from two major drawbacks. First they were highly lipid soluble (log P > 2) and were able to readily penetrate the blood brain barrier (BBB), and their interaction with central H_1 receptors invariably led to side effects such as sedation and psychomotor impairment. In addition, the agents tended to be non-selective for H_1 receptors at the doses given and exhibited anticholinergic (muscarinic) and anti-adrenergic activities adding to the side effect profile. The search for more potent, H_1-selective and hydrophilic (to prevent transport across the BBB) agents eventually led to terfenadine **1** (and, more recently, its metabolite fexofenadine **2**), loratidine **3**, cetirizine **4** and astemizole **5** which are now the most prescribed drugs for seasonal rhinitis.

8.2 Terfenadine and Fexofenadine

Terfenadine was one of the first second generation H_1 blockers that was free from anticholinergic and CNS-related side effects and was marketed as the racemate. The compound undergoes almost complete first pass metabolism

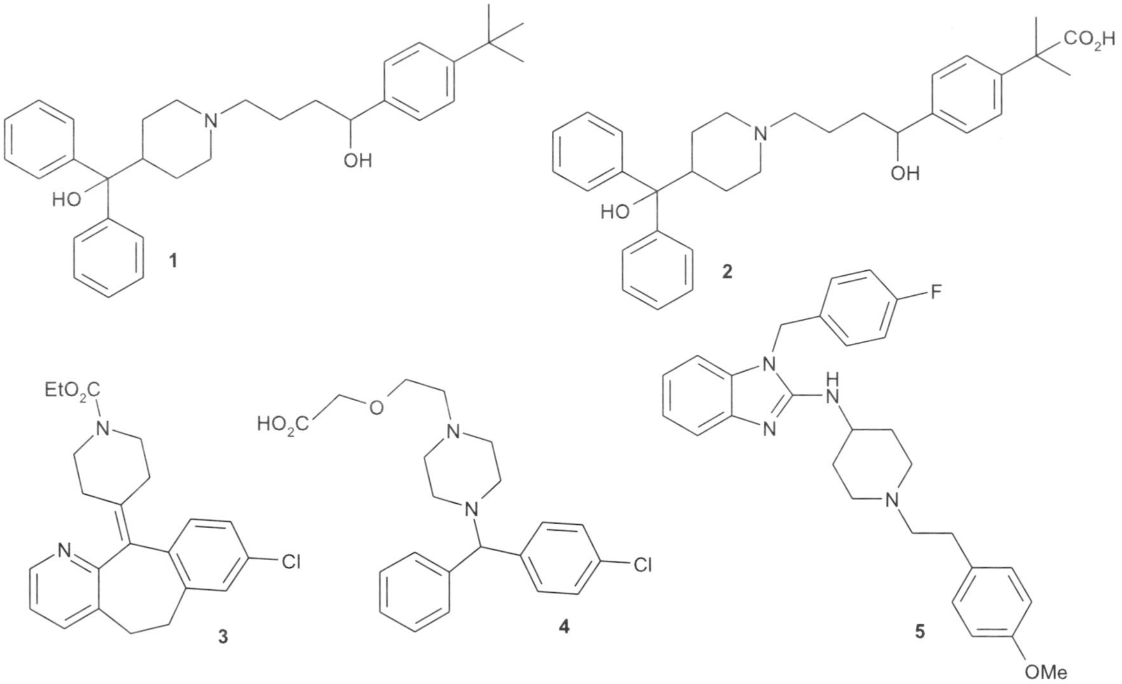

Scheme 8.1.   Second generation antihistamines.

(99%) to the pharmacologically active acid **2** (fexofenadine; formally, terfenadine is a pro-drug of fexofenadine) and the inactive N-dealkylated piperidine **6** by the action of hepatic cytochrome P_{450} enzymes. Under certain circumstances, such as hepatic impairment or when co-administered with other drugs which compete for the oxidative metabolism machinery (macrolide antibiotics, antifungal agents), significant concentrations of unchanged terfenadine appear in the plasma. This has been implicated with serious cardiac adverse events resulting in the withdrawal of the compound from the market place during 1998. Fortunately, fexofenadine itself is rapidly absorbed when given orally, does not build up in hepatically impaired patients and is devoid of the cardiac complications associated with terfenadine.

The 4-pyridine methanol **7** was available from isonicotinic acid ethyl ester **8** by a double Grignard reaction with phenyl magnesium bromide and this material as the HCl salt was then hydrogenated in the presence of PtO_2. Azacyclonol **6** was alkylated with the chlorobutanone **9** in refluxing toluene in the presence of an inorganic base and a catalytic amount of NaI. To complete the synthesis of **1**, the ketone group in **10** was reduced with KBH_4 to the racemic alcohol.

The Freidel Crafts's reaction shown above to give **9** (Scheme 8.2) gives a 2:1 mixture of the meta and para acylated products when ethyl dimethylphenylacetate is used as the substrate and hence is not convenient for the synthesis of fexofenadine. To overcome this lack of regioselectivity, a palladium catalysed coupling of the arylbromide **11** with 3-butyn-1-ol (Scheme 8.3) was

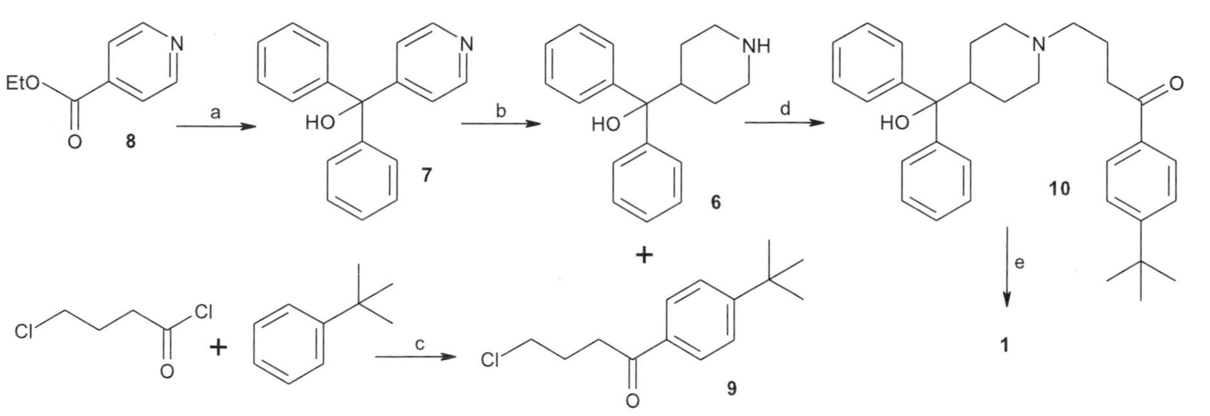

Scheme 8.2. Reagents: (a) PhMgBr, Et$_2$O, −20°, 45 min then reflux 1 h; (b) PtO$_2$, MeOH, H$_2$, 3–4 atm.; (c) AlCl$_3$; (d) KHCO$_3$, KI, PhMe, reflux, 60 h; (e) KOH, MeOH, KBH$_4$, 0°, 1 h.

employed. 4-Bromophenylacetic acid was first esterified with methanol in the presence of Me$_3$SiCl and then doubly methylated at the benzylic position to afford **11**. After the Pd(0) catalysed coupling, the alcohol **12** was activated for nucleophilic displacement as the mesylate and then reacted with azacyclonol. Acid catalysed hydration of the alkyne **13** in the presence of mercuric oxide gave **14** which was reduced to the alcohol with NaBH$_4$. Saponification of the ester yielded fexofenadine **2**.

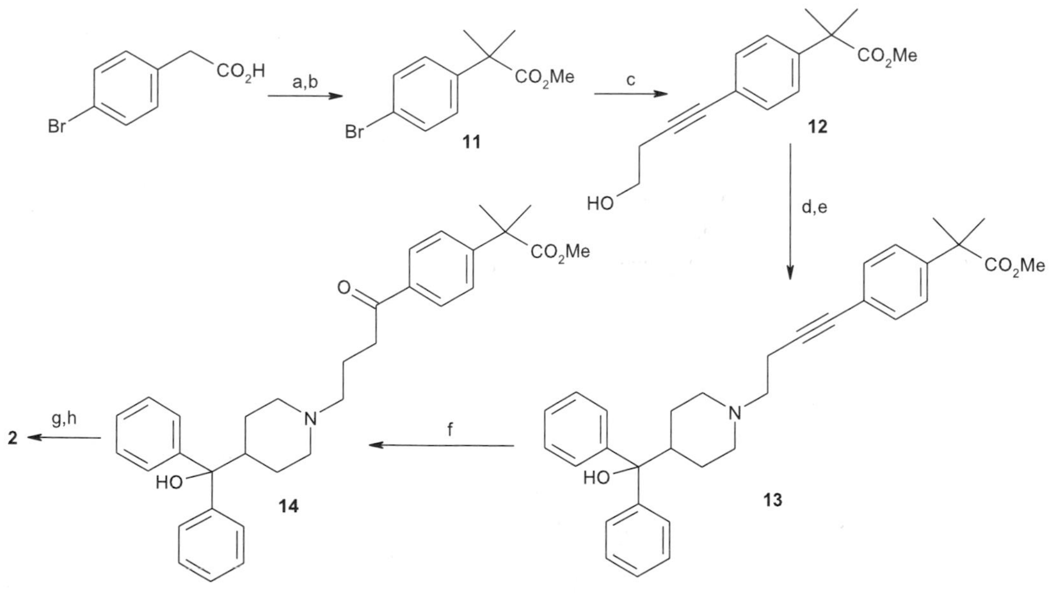

Scheme 8.3. Reagents: (a) Me$_3$SiCl, MeOH, rt, 15 h; (b) THF, NaH, MeI (2.2 equiv.), 15 h, rt; (c) (Ph$_3$P)$_4$Pd(0), CuBr, 3-butyn-1-ol, Et$_3$N, reflux 3.5 h; (d) MeSO$_2$Cl, DCM, py, 20 h, rt; (e) azacyclonol.HCl, K$_2$CO$_3$, MeCN, reflux, 12 h; (f) HgO, H$_2$SO$_4$, MeOH, 55°, 3 h; (g) NaBH$_4$, MeOH, 20 h, rt; (h) KOH, H$_2$O, MeOH, 16 h, 80°.

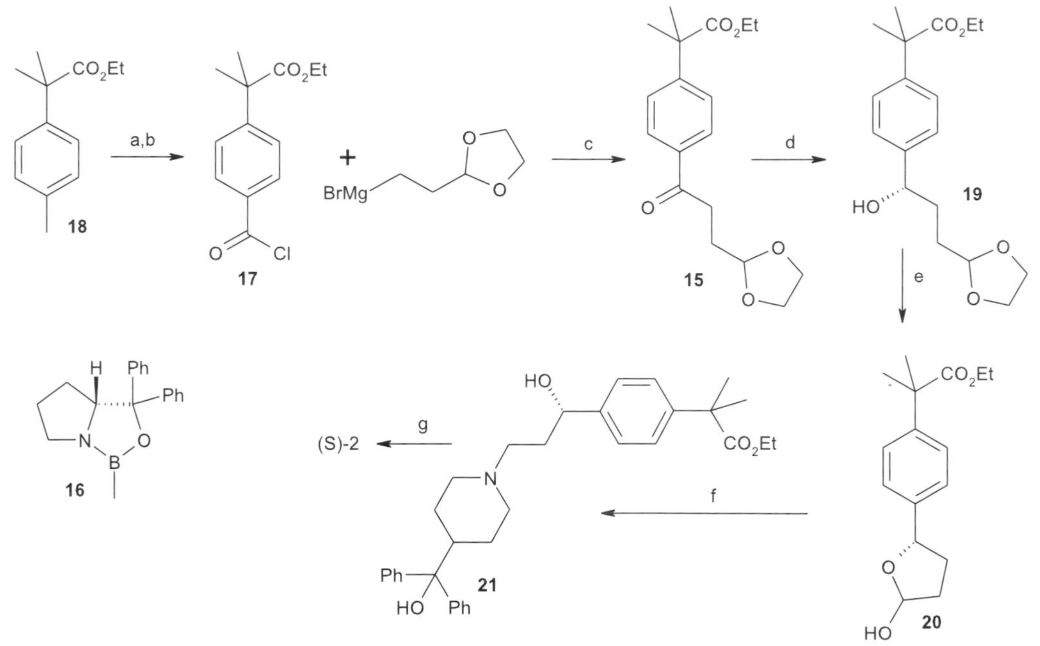

Scheme 8.4. Reagents: (a) KMnO$_4$, pyridine; (b) (COCl)$_2$; (c) cat. Fe(acac)$_3$, THF; (d) BH$_3$:Me$_2$S, 16, −10°, 15 min; (e) Amberlyst-15, Me$_2$CO, H$_2$O, 50°; (f) MeOH, 50°, NaBH$_4$; (g) 2N NaOH, MeOH.

Resolution of the fexofenadine ethyl ester as its (+)-di-para-toluoyltartaric acid salt allowed the absolute stereochemistry of the (−)form to be assigned as (S). This enantiomer could be prepared by an asymmetric synthesis wherein the prochiral ketone **15** (Scheme 8.4) was reduced with diborane in the presence of the oxazaborolidine catalyst **16**. Thus, the acid chloride **17** was prepared from the corresponding toluene **18** by permanganate oxidation followed by exposure to oxalyl chloride and this reacted smoothly in a Fe(acac)$_3$ catalysed Grignard reaction to afford the ketone **15**. Asymmetric reduction of the ketone gave the (S)-alcohol **19** which was deprotected under mild acid conditions. The hydrolysis step proceeded with concomitant ring closure to the lactol **20** without epimerisation of the stereogenic center as assessed by chiral HPLC (96% ee) and by specific rotation of the ester **21** of the final product. Reductive alkyation of azacyclonol with the lactol followed by saponification of the ethyl ester gave **(S)-2**.

8.3 Loratadine

Azatadine (des-chloro-**22,** Scheme 8.4) is a potent antihistamine but is able to penetrate the BBB and hence has sedative activity quite like the first generation compounds. Using variously substituted derivatives of **22** as starting points to discover new compounds devoid of the central action, it was discovered that the ethylcarbamate, loratadine **3**, had this property whilst retaining potent H$_1$ blocking activity. Interestingly, however, the compound is

Scheme 8.4. Reagents: (a) diethyl 3-chlorobenzylphosphonate, NaOMe, DMF, 35°, then 0.5 h at rt; (b) 5% Pd-C, EtOAc; (c) HOAc, H_2O_2, 20 h, 60°; (d) DMS, 80°, 3 h; (e) H_2O, NaCN, 20 h, rt; (f) NaOH, EtOH, reflux; (g) $SOCl_2$, PhH, reflux, 1.5 h; (h) CS_2, $AlCl_3$, 19 h, rt; (i) Et_2O, Mg, N-methyl-4-chloropiperidine, reflux 6 h; (j) c.H_2SO_4, 2 h, rt; (k) ethyl chloroformate PhH, reflux, 2 h then 16 h at rt

rapidly metabolised to the secondary amine by hydrolysis and decarboxylation by P_{450} enzymes in the liver. This metabolite has a greater half life than the parent and presumably is the major contributor to the beneficial drug effect.

The original route to loratidine involved demethylation and carbamoylation of **22** which had first been prepared as part of structure activity studies of azatadine. Thus it followed closely the synthesis of azatadine but suffered from the reactivity (towards reduction) and asymmetry (during the cyclisation step) imposed by the chloro group (Scheme 8.4). A Wadsworth-Emmons reaction with diethyl 3-chlororbenzylphosphonate on pyridine-3-carboxalde-hyde gave the *trans*-stilbazole **23** which was hydrogenated and converted to the N-oxide **24**. O-Methylation of the N-oxide with dimethylsulphate followed by displacement with cyanide gave the 2-cyanopyridine **25** in admixture with the 6-cyano isomer which had to be removed by careful fractional distillation. Hydrolysis to the acid and activation of **26** as the acid chloride gave the intermediate necessary for the Friedel Craft's cyclisation reaction and this proceeded to yield a mixture of isomers (4:1 in favour of the 8-chloro isomer) from which desired material **27** was isolated by crystallisation. The ketone was reacted with the Grignard reagent from 1-methyl-4-chloropiperidine and the resulting benzylic alcohol eliminated under acidic conditions to form **22**.

Scheme 8.5. Reagents: (a) t-BuOH, 70°, c. H₂SO₄; (b) 2 equivalents of n-BuLi, THF, –40°; (c) POCl₃, heat, 3 h; (d) Et₂O, Mg, 1-methyl-4-chloropiperidine, add nitrile, 50° 1 h, then 2 N HCl, 2 h, rt; (e) HF, BF₃, –35°; (d) PhCH₃, ClCO₂Et, 80°, 1 h.

Using ethyl chloroformate, the basic nitrogen of **22** was quaternised and demethylated in situ to yield loratidine.

To overcome the problems described above, an alternative procedure started with 2-cyano-3-methylpyridine **28** (Scheme 8.5) which was converted by a Ritter reaction to the corresponding amide **29** with t-BuOH and acid. The dianion of **29** was generated with two equivalents of n-butyl lithium and regioselectively alkylated on the 3-methyl group with 3-chlorobenzyl chloride to afford **30.** Dehydration of the amide with POCl₃ (to the nitrile **25**) which could be taken on to loratidine by the original route. However, it was found that direct Grignard reaction on the nitrile and *in situ* hydrolysis of the imine gave the ketone **31.** Cyclodehydration was most conveniently performed by superacid catalysis, the preferred reagent being HF/BF₃, and the overall process was suitable for multi-kilogram scale synthesis.

8.4   Cetirizine

Cetirizine was first prepared and is sold as the racemic mixture (Scheme 8.6). The asymmetric diphenylmethane **32** can be made by several methods, most conveniently by alkylation of piperazine with 4-chlorobenzhydryl chloride **33** and chromatographic purification of the mixture of products. This intermediate was alkylated at the remaining secondary amine with methyl 2-chloroethoxyacetate and the resulting methyl ester hydrolysed under basic conditions to yield cetirizine.

Separation of the enantiomers of cetirizine by classical resolution of the tartrate salts was low yielding and afforded poor enrichment. However, it had

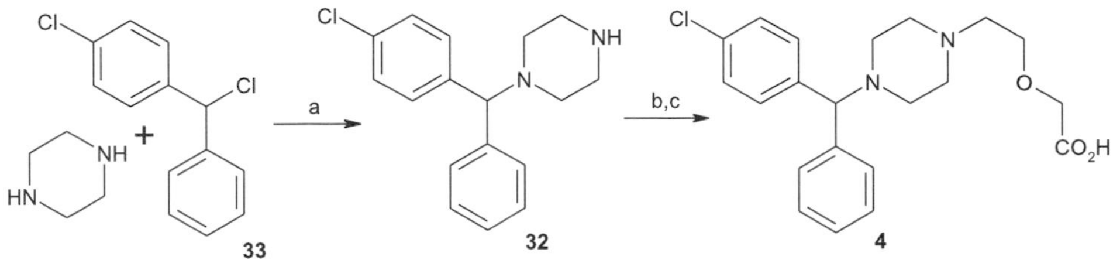

Scheme 8.6. Reagents: (a) dioxane, 7 h, reflux; (b) Na$_2$CO$_3$, xylene, 40 h, reflux; (c) 1N KOH in EtOH, reflux, 1 h.

been established that the diarylmethylamine **34** (Scheme 8.7) could be resolved efficiently by such methods and this was used as the key intermediate in synthesis of (+)- and (−)-cetirizine on the kilogram scale. Starting from the benzophenone **35**, reaction with ammonium formate gave the N-formyl derivative **36** which was hydrolysed with aqueous acid to the amine. Resolution with either (+)- or (−)- tartaric acid followed by basification of the separated salts afforded both (+)- and (−)-amines. After considerable experimentation, it was found that the N-tosyl bis-alkylating agent **37** in reaction with (−)-**34** gave the desired piperazine (−)-**38** after deprotection under strongly acidic conditions optimised to avoid epimerisation (> 99% ee by chiral HPLC on a kilogram scale). Most conveniently, it was discovered that the reductive removal of the tosyl group was best achieved with HBr in the presence of a phenol such as 4-hydroxybenzoic acid which could be easily removed during workup. N-alkylation of the secondary amine under Finklestein conditions gave (−)-**39**, which, after hydrolysis of the primary amide, afforded the dextrorotatory isomer of the desired acid **4**.

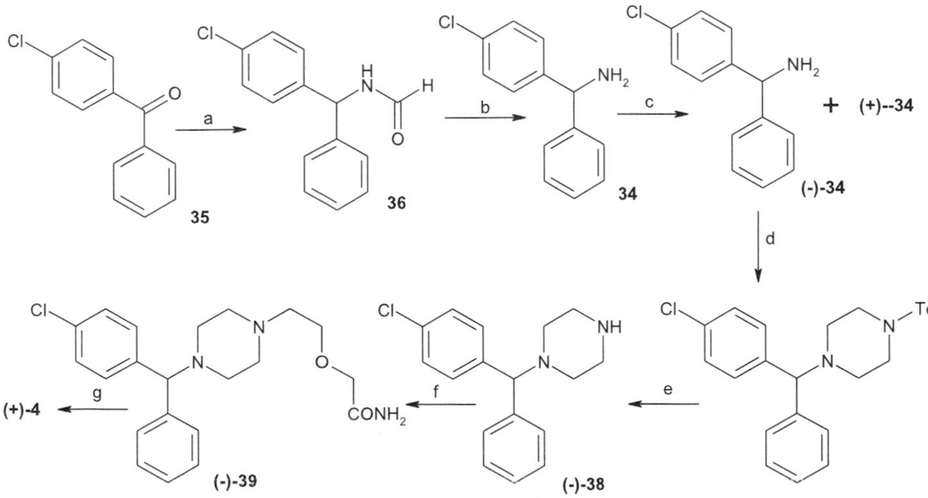

Scheme 8.7. Reagents: (a) HCO$_2$NH$_4$, heat at 180°; (b) 10% HCl, reflux; (c) (+)- or (−)-tartaric acid, recrystallise from ethanol then take separated enantiomers forward; (d) TosN(CH$_2$CH$_2$Cl)$_2$ (37), iPr$_2$NEt, reflux, 3.5 h; (e) 30% HBr, AcOH, p-HOC$_6$H$_4$CO$_2$H, rt, 2 days; (f) Cl(CH$_2$)$_2$OCH$_2$CONH$_2$, Na$_2$CO$_3$, KI, PhMe; (g) HCl, H$_2$O, 50°.

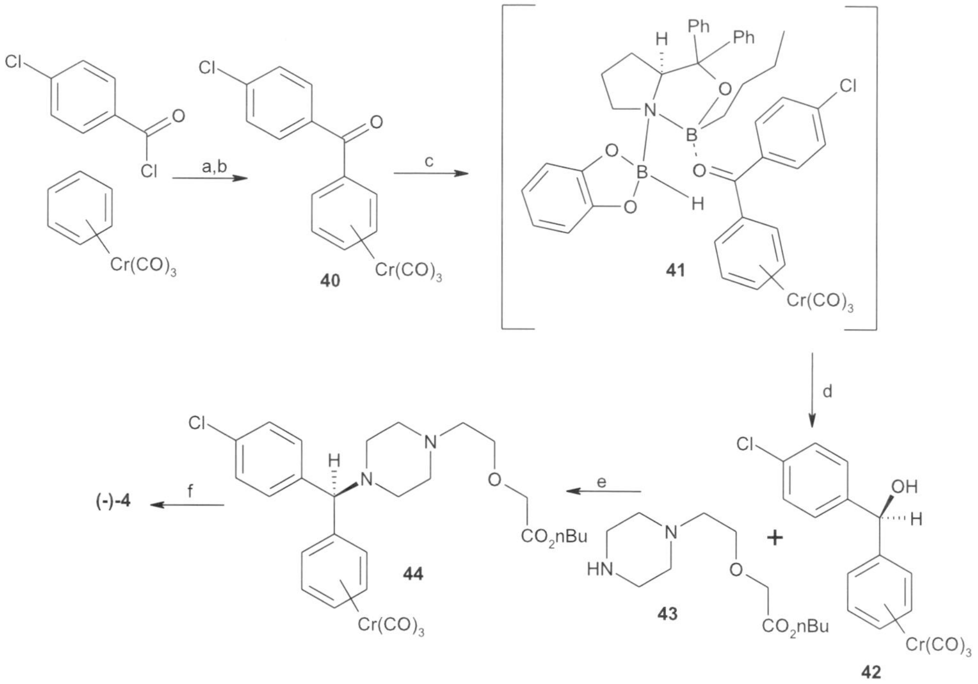

Scheme 8.8. Reagents: (a) n-BuLi, TMEDA, THF, −78°, 45 min., then CuBr, Me₂S, 30 min; (b) 4-Cl-benzoyl chloride, −78°, then rt, 18 h; (c) catecholborane, oxaborolidine catalyst, PhMe, −78° to −40°, 2 h; (d) HBF₄:Et₂O, DCM, 1 min, −60°; (e) add **43**; (f) pyridine, reflux, 4 h; (g) 2 M HCl, 50°, 4 h.

The first enantioselective synthesis combined two key observations: (a) π-Cr(CO)$_3$ complexation of one aryl ring in a benzophenone produces 'electronic dissymmetry' and (b) chiral oxazaborolidine reduction of such complexes may be performed enantioselectively. Metalation of benzene chromium tricarbonyl followed by acylation with 4-chlorobenzoyl chloride gave the benzophenone **40**. This was reduced by catecholborane catalysed by the chiral oxaborolidine reagent wherein the complex formed (**41**) has the oxaborolidine B-atom complexed to the ketone oxygen adjacent to the less bulky ketone substituent – in this case the chlorophenyl moiety. The oxaborolidine was destroyed using tetrafluoroboric acid and the resulting secondary alcohol **42** reacted with the mono-substituted piperazine **43** with complete retention of configuration at the benzylic center. In this context, the chromium complex preformed its second role, that is to serve as a 'stereocontroller' during the displacement. The chromium in **44** was removed by heating with pyridine and ester hydrolysis under acidic conditions afforded (−)-cetirizine.

8.5 Astemizole

Astemizole was the result of a detailed investigation of antihistamine properties in a series of benzimidazoles which, although at the time was unrelated structurally to known antihistamines, were deemed to have the pharmacophore responsible for this activity – a basic nitrogen separated by

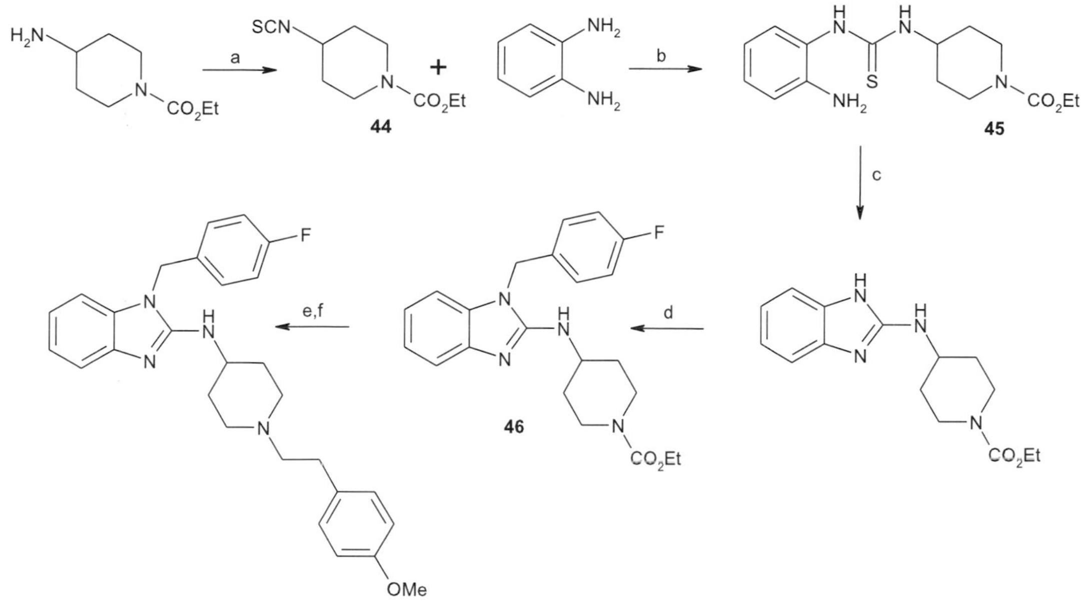

Scheme 8.9. Reagents: (a) NaOH, H$_2$O, CS$_2$, ClCO$_2$Et, 10° then 60°, 2 h; (b) EtOH, rt, 16 h; (c) MeI, EtOH, reflux 8 h; (d) p-F-C$_6$H$_4$CH$_2$Br, Na$_2$CO$_3$, DMF, 70°; (e) 48% HBr, reflux 3 h; (f) p-MeO-C$_6$H$_4$(CH$_2$)$_2$OMs, Na$_2$CO$_3$, DMF, 16 h, 70°.

4.1–4.2 Angstroms from a C or N atom bearing two lipophilic entities (see Scheme 8.1). The isothiocyanate **44** was reacted with o-phenylenediamine in refluxing ethanol to give the urea **45** ready for cyclodesulphurisation which could be mediated by a variety of reagents. Initially, MeI was used to generate the reactive -SMe intermediate which cyclised under the reaction conditions. Subsequently HgO-mediated cyclisation was preferred. It was then necessary to selectively alkylate the *endo*-N atom of the aminobenzimidazole with p-fluorobenzyl bromide and this was achieved under mildly basic conditions using sodium carbonate. Deprotection of **46** with aqueous HBr followed by alkylation at the more basic piperidine nitrogen gave astemizole.

A more recent approach to the intermediate **46** involves a Pd-catalysed amination reaction (Scheme 8.10) based on the Buchwald and Hartwig

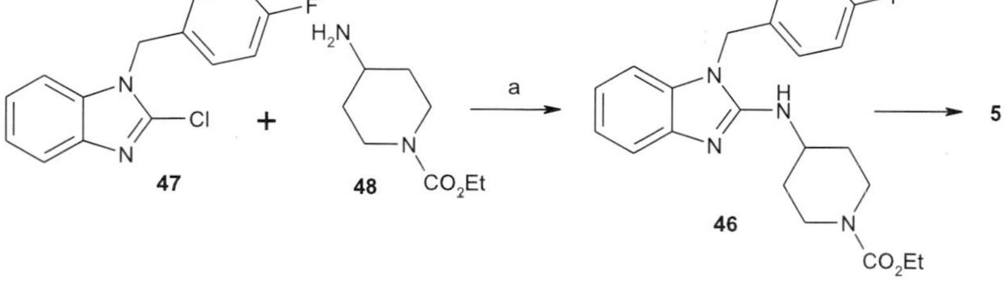

Scheme 8.10. Reagents: (a) NaOBu-t, Pd$_2$(dba)$_3$, BINAP, PhMe, 85°, 1 h.

procedures for amination of aryl bromides. In this instance, the N-substituted 2-chlorobenzimidazole **47** was treated with the amine **48** and the reaction outcome was highly dependent on the nature of the Pd coordination sphere and the added ligand. Optimally, $Pd_2(dba)_3$ with BINAP using sodium t-butoxide as base was used to obtain a yield of > 98%. In the absence of the catalyst less than 2% conversion to **46** was observed over a 24 hour period in refluxing toluene.

References:

Terfenadine and Fexofenadine
A. A. Carr and R. Kinsolving, US Patent, 1975, 3,878,217
A. A. Carr and D. P. Meyer, *Arzneim Forsch,* 1982, **32**, 1157
S. H. Kawai, R. J. Hambalek and G. Just, *J. Org. Chem.*, 1994, **59**, 2620
Q. K. Fang et al., *Tet. Lett.*, 1998, **39**, 2701.

Loratidine
F. J. Villani et al, US Patent, 1981, 4,282,233
F. J. Villani, et al., *Arzneim Forsch*, 1986, **36**, 1311
F. J. Villani et al., *J. Het. Chem.*, 1971, **8**, 73
D. P. Schumacher et al., *J. Org. Chem.*, 1989, **54**, 2242

Cetirizine:
E. Baltes, J. de Lannoy and L. Rodriguez, US Patent, 1985, 4,525,358
C. J. Opalka et al., *Synthesis*, 1995, 766
E. J. Corey and C. J. Helal, *Tet Lett* 1996, **37**, 4837

Astemizole
F. Janssens et al., US Patent, 1980, 4,219,559
F. Janssens et al., *J. Medicin. Chem.*, 1985, **28**, 1934
Y. Hong et al., *Tet. Lett.*, 1997, **38**, 5607

9 Nucleoside anaologues which inhibit HIV reverse transcriptase as anti-AIDS drugs

9.1 Introduction

HIV is the causative agent of AIDS, which is defined as the development of certain opportunistic infections, cancers and encephalopathy and is associated with a catastrophic depletion of CD4$^+$ lymphocytes. It is not yet understood how HIV causes AIDS and the pathogenic mechanisms underlying the HIV disease are complex. Upon infection of CD4$^+$ cells, the reverse transcribed RNA genome of HIV is integrated into the host genome. As the genome is integrated and HIV disease is chronic, and since there are no strategies for the excision of viral genomes from the host DNA, it is probable that treatment of HIV infected individuals by anti-viral agents will be lifelong.

Like other retroviruses, HIV carries its genetic information as (+)-strand RNA, two copies of which are contained within the virion protein core along with virally encoded enzymes responsible for transcribing the genetic information into DNA once inside the host cell. Pivotal in the whole replicative process is the role of the HIV enzyme, reverse transcriptase (RT). This enzyme is multifunctional having both RNA-dependent DNA polymerase and DNA-dependent DNA polymerase activities as well as an inherent RNase H activity. Following formation of the (–)-DNA strand, it is the RNase H activity that is necessary to degrade the original viral RNA of the newly formed RNA-DNA heteroduplex so that the double stranded DNA can be prepared ready for integration into the host cell DNA. Nucleoside analogues which are active inhibitors of HIV replication act, after phosphorylation by host cell kinases to their respective 5′-triphosphates, as competitive inhibitors and/or chain terminators of HIV-RT. Typically, chain terminators are devoid of the crucial 3′-OH group necessary for continued chain extension.

The first drug shown to have widespread utility in the management of AIDS was zidovudine (**1**; AZT). The compound had been prepared in 1964 as a potential anti-cancer agent but was shown to be ineffective and was subsequently re-screened as an antiviral as the AIDS epidemic hit Western society during the mid-1980's. It is still the central component in the multitude of combination therapies now approved or still being evaluated clinically. As HIV is highly variable – it is an RNA virus with no proofreading function on the RT – it is inevitable that drug resistant strains will emerge over long term treatment and hence there is a requirement for a succession of new compounds to be developed. One such compound that has been found to be of immense value in combination with AZT is lamivudine ((−)-*cis*-2; 3-TC).

9.2 Zidovudine (AZT)

Interestingly, AZT was first prepared not as a biologically active molecule but as an intermediate towards 3′-amino nucleotides (i.e. as the monophosphate) in turn projected to act as active site directed inhibitors of nuclear exoribonuclease of unspecified therapeutic utility. The primary alcohol of (2′-deoxy-lyxofuranosyl)thymine **3** (Scheme 9.1) was selectively protected with trityl chloride and the remaining secondary alcohol activated for nucleophilic displacement as the mesyl derivative **4**. Treatment of **4** with lithium azide led to **5** by inversion of configuration which deprotected smoothly under acidic conditions.

In a second approach (Scheme 9.2), thymidine **6** acted as the starting material which was again protected as the trityl derivative on the 5-OH and then mesylated to give **7**, the epimer of the intermediate **4** described above. In the presence of potassium phthalimide, elimination of sulphonic acid took place to generate the key intermediate anhydro sugar **8**. Ring opening with sodium azide followed by deprotection gave **1**. Subsequently it was found that the unprotected anhydro sugar **9** was available directly from thymidine by displacement of the 3′-OH in the presence of chloropentafluorotriethylamine.

Alternative approaches were investigated to avoid the use of the expensive thymidine starting material; D-mannitol is a readily available alternative which can be converted into the D-glyceraldehyde derivative **10** (Scheme 9.3) in two steps. A stabilised Wittig reaction introduced the two remaining C-atoms of the developing ribose ring with **11** being produced as an 8:1 mixture with the

Scheme 9.1. Reagents: (a) Tr-Cl, pyridine; (b) pyridine, MeSO$_2$Cl, 0°, 16 h; (c) LiN$_3$, DMF, 100°, 3 h; (d) CHCl$_3$, HCl, 1 h, 4°.

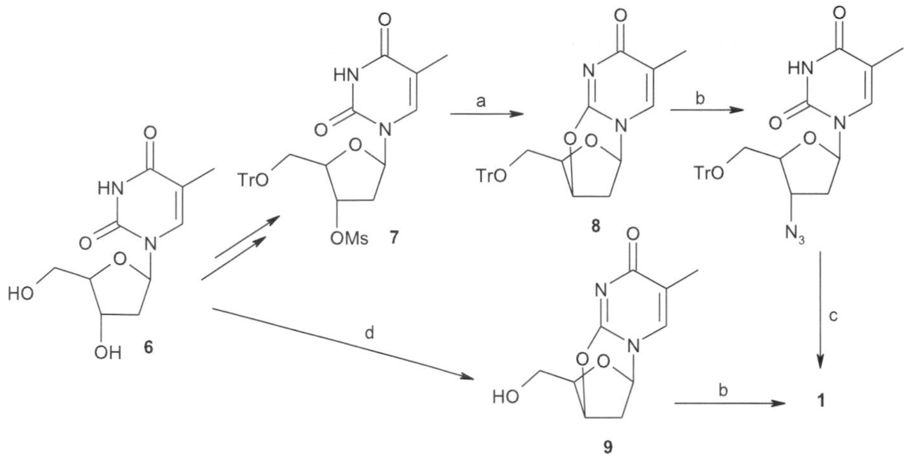

Scheme 9.2. Reagents: (a) K-Phthalimide, DMF, H_2O, 95°, 20 min; (b) DMF, NaN_3, H_2O, reflux, 11 h; (c) 80% HOAc, 1.75 h, steam bath; (d) chlororpentafluorotriethylamine.

E-isomer which was removed chromatographically. Acid treatment of **11** gave the lactone **12** and this, in protected form, underwent a Michael addition with azide with no sign of the β-adduct since this face is shielded by the bulky 5′-protecting group. Reduction at low temperature to the lactol and activation of the anomeric hydroxyl group as the acetoxy derivative **13** gave the sugar intermediate required for condensation with silylated thymine under Vorbruggen conditions. This reaction proceeded with no stereochemical control giving a 1:1 mixture of α- and β-anomers **14** which could only be separated cleanly after removal of the silyl protecting group in the final product **1**.

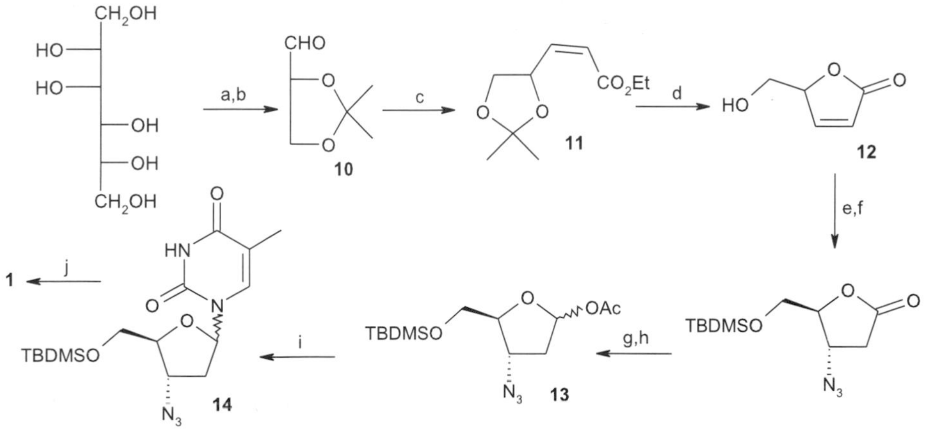

Scheme 9.3. Reagents: (a) Me_2CO, H+, reflux; (b) $Pb(OAc)_4$; (c) $Ph_3P{=}CHCO_2Et$, MeOH, 0°; (d) HCl; (e) t-Bu(Me)$_2$SiCl, imidazole, DMF; (f) LiN_3, THF, AcOH, H_2O; (g) DIBAL, DCM, −78°; (h) Ac_2O, pyridine; (i) di-TMS-thymine, TMS-triflate, EDC; (j) n-Bu$_4$N+F$^-$, THF.

Scheme 9.4.   Reagents: (a) HC(OMe)$_3$, DCM, ZnCl$_2$, 25°, 18 h.; (b) DIBAL, −78°, then warm to 25°, 10 h; (c) D-(−)-diisopropyl tartrate, DCM, −20°, Ti(O-iPr)$_4$, 15 min then add 16, t-BuOOH, store at −20°, 2 days; (d) TMSN$_3$, DCM, 0°, Et$_2$AlF, then 25°, 2 days; (e) 1.5% HCl, DCM, 5 min; (f) DMF, imidazole, t-BuPh$_2$SiCl, 25°, 3 h; (g) TMS triflate, EDC.

Although somewhat lengthy, another method based on an enantioselective Sharpless epoxidation proved successful. Crotonaldehyde was converted into a mixture of (E)- and (Z)-trimethylsilyloxybutadiene (Scheme 9.4) and treated with orthoformate in the presence of a Lewis acid to give the enal **15**. Reduction with DIBAL gave **16** as a 95:5 mixture of (E)- and (Z)-geometric isomers which was subjected to the asymmetric epoxidation conditions of Sharpless. The resulting epoxy alcohol **17** was opened with azide in the presence of diethylaluminium fluoride to give the (2S,3S) azide **18**. Acid catalysed ring closure in dilute solution to suppress hexose ring formation gave predominantly the desired furanose **19** which could be purified chromatographically. After protection of the 5′-OH to give **20**, and using again Vorbruggen conditions, the anomeric methoxy group was displaced by silylated thymine. From the mixture of anomers so obtained, the desired β-anomer **1** was isolated by flash chromatography.

9.3   Lamivudine (3-TC)

BCH-189, the *cis*-racemate from which 3TC ((−)-*cis*-**2**) is derived, was first reported in the summer of 1989; it is the β-anomer with *cis*-stereochemistry around the oxathiolane ring which displays antiviral activity *in vitro* comparable to AZT. Early approaches to the synthesis of BCH-189 utilised the intermediacy of oxathiolane **17** obtained as a 1:1 mixture of anomers from reaction of benzoyloxyacetaldehyde **15** with mercapto-acetaldehyde **16** (Scheme 9.5). Cytosine, the base required for the coupling reaction was silylated with hexamethyldisilane, and the disilylated material **18** reacted under Lewis acid catalysis (TMS-triflate) to afford a 1:1 mixture of the *cis*- and *trans*-oxathiolane **2** but only under forcing conditions and with low yield

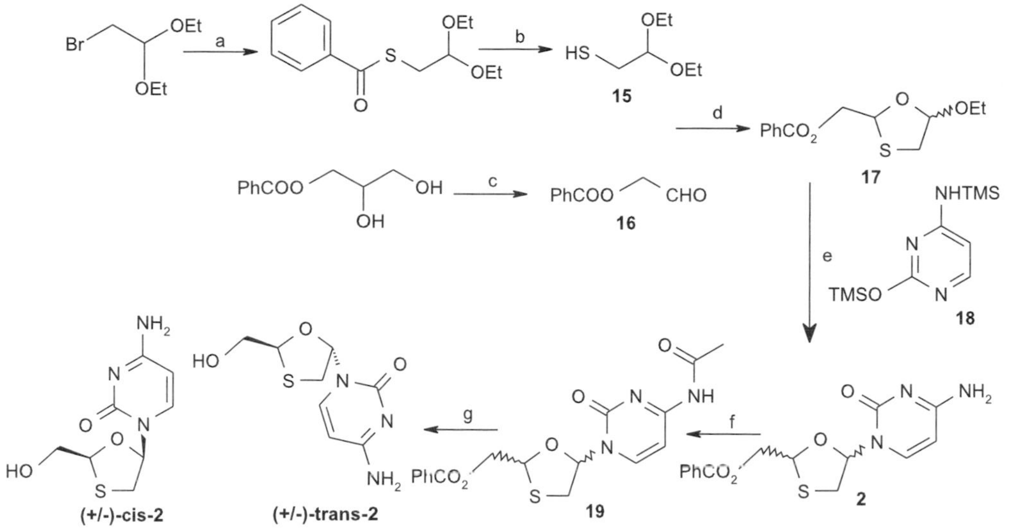

Scheme 9.5. Reagents: (a) KOt-Bu, DMF, PhCOSH; (b) THF, NaOH, H_2O, 15 h, reflux; (c) DCM, H_2O, $NaIO_4$, rt, 2 h; (d) pTSA, PhMe, 120°, distill off EtOH; (e) TMSOTf, MeCN, reflux 3 days; (f) Ac_2O, DMAP, pyridine, 16 h, rt then chromatography to isolated pure *cis*- and *trans*-19; (g) NH_3, MeOH, rt, 16 h.

(30%). To effect separation of the geometric isomers, **2** was acetylated with Ac_2O and the product **19** chromatographed with the pure *trans*-racemate eluting first from the column. Deprotection of the now separated **19** gave *cis*- and *trans*-**2**.

To correct two of the deficiencies of the original route, both the leaving group and the Lewis acid used in the coupling reaction were investigated. The yield of the coupling reaction was improved significantly by replacing the leaving group ethoxy with acetoxy (Scheme 9.6). Once it was realised that

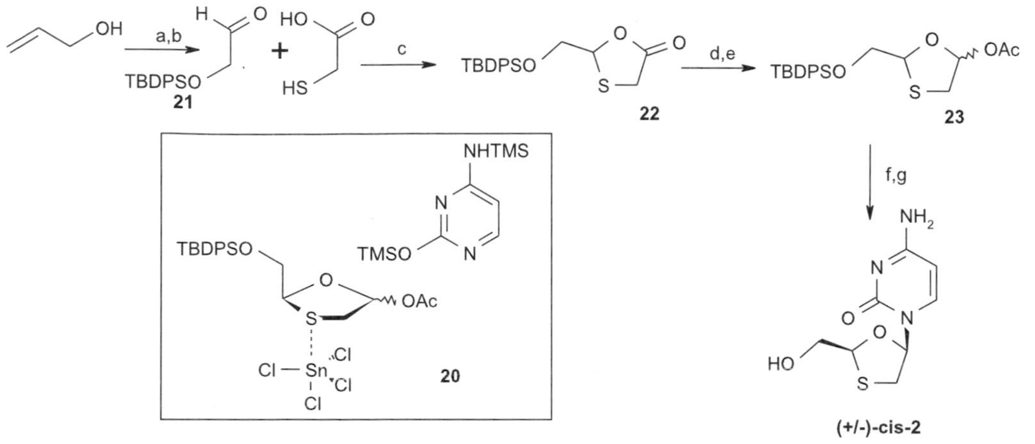

Scheme 9.6. Reagents: (a) NaH, THF, TBDPS-Cl, 0°, 1 h; (b) DCM, O_3, −78° then DMS, −78° to rt, 16 h; (c) PhMe, reflux 2 h under Dean & Stark conditions; (d) DIBAL, PhMe, 30 min, −78°; (e) Ac_2O, rt, 16 h; (f) DCM, di-TMS-cytosine, $SnCl_4$, rt; (g) TBAF.

antiviral activity resided in the *cis*-racemate, various alternative coupling reactions other than the standard Vorbruggen conditions were studied to optimise the proportion of *cis* being formed. It was shown that the use of SnCl$_4$ in place of trimethylsilyltriflate afforded almost exclusively the β-anomer (> 300:1 by HPLC and X-ray crystallography on the deprotected nucleoside) possibly by complex formation between the oxathiolane ring and SnCl$_4$ from the less hindered lower face forcing the incoming silylated base to approach from above **20**. Thus, allyl alcohol was protected and ozonised to give, after a reductive work up, the aldehyde **21**. This was condensed with mercaptoacetic acid, the resulting lactone **22** reduced and the product acetylated to afford **23** ready for N-glycosylation of silylated cytosine.

Resolution of BCH-189, first in milligram amounts by preparative HPLC and subsequently in gram quantities by enzymatic methods, showed unexpectedly that both *cis*-enantiomers had potent antiviral activity. The first enzymatic process (Scheme 9.7) relied upon recognition of the 'natural' enantiomer of BCH189 monophosphate by the 5'-nucleotidase from *Crotalus atrox* venom leaving the 'unnatural' enantiomer as the unchanged monophosphate. Facile separation of these two products and subsequent dephosphorylation using bacterial alkaline phosphatase of promiscuous specificity afforded the separated isomers. Once it was appreciated that the 'unnatural' isomer was the one of real interest, a more direct method using cytidine deaminase (Scheme 4b) from *E. coli* to degrade the natural isomer to the uridine analogue was employed. Again, ready separation of the resulting mixture afforded (–)-BCH-189, 3-TC in greater than 98% ee. Cloning and overexpression of the enzyme under the control of a high level inducible promotor and immobilisation of the enzyme on Eupergit-C to allow it's reuse over many cycles has generated a multikilogram synthesis of essentially enantiomerically pure 3-TC.

Scheme 9.7. Reagents: (a) POCl$_3$, (MeO)$_3$PO; (b) 5'-ribonucleotide phosphohydrolase from *Crotalus atrox;* (c) separate on DEAE sephadex; (d) bacterial alkaline phosphatase.

Scheme 9.8. Reagents: (a) BF$_3$:Et$_2$O, MeCN; (b) chromatography on SiO$_2$; (c) Pb(OAc)$_4$, DMF; (d) silylated cytosine, TMSI, DCE then separate cis and trans isomers; (e) Amberlite IRA400(OH), EtOH.

The first enantioselective synthesis of relied upon the preparation of homochiral oxathiolane intermediates which may be coupled to the cytosine base under conditions which preserve stereochemical integrity. From the available chiral pool, (+)-thiolactic acid **24** (Scheme 9.8) was condensed with 2-benzoyloxyacetaldehyde to give a 1:2 mixture of diastereomeric oxathiolane acids **25a** and **25b**. Separation and oxidative decarboxylation of **25b** afforded an anomeric mixture of *anti* and *syn* acetates which were coupled with silylated cytosine in the presence of iodotrimethylsilane to give a 1.3:1 mixture of **26a** and **26b**. Chromatographic separation of the diastereomers and deprotection yielded 3-TC. The overall efficiency of this process is, however, limited by failure to control the stereochemistry of the base coupling reaction. Coupling in the presence of SnCl$_4$, which had previously been shown to offer such control, gave a racemic product possibly because opening and closing of the oxathiolane ring was rapid under these conditions.

Not surprisingly, other groups investigated synthetic methods which take advantage of stereochemistry built into carbohydrates. Given that it had been anticipated that antiviral activity would reside in the 'natural' isomer, the (+)-enantiomer of 3-TC had already been made by a multistep process from the cheap and readily available D-mannose; L-mannose is too expensive to provide a practical approach to the (−)-enantiomer. Instead, L-gulose **27**, prepared in two steps from L-ascorbic acid (Scheme 9.9), was tosylated on the primary alcohol and then per-acetylated to give **28**. Displacement of the anomeric acetoxy group by bromine gave **29** which was reacted with potassium O-ethylxanthate to generate, after deprotection, the thioanhydro-L-gulose **30**. Next the *cis*-diol group was reductively cleaved with periodate

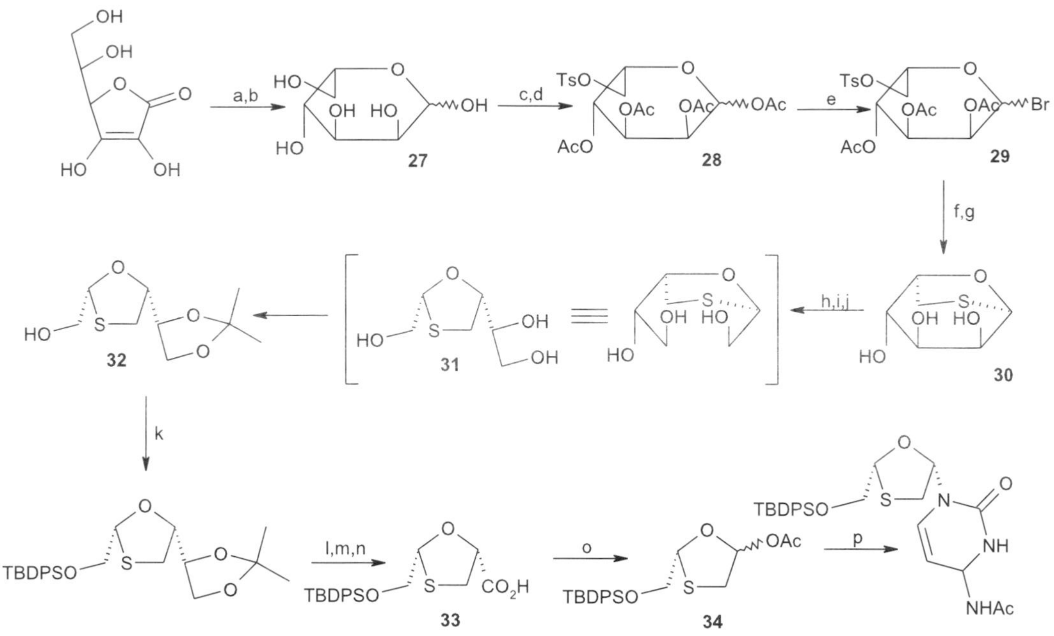

Scheme 9.9. Reagents: (a) Pd-C, H$_2$, 50° H$_2$O;(b) NaBH$_4$; (c) TsCl; (d) Ac$_2$O, pyridine, 20°; (e) HBr, AcOH, 0°; (f) EtOC(=S)SK, Me$_2$CO, reflux; (g) NH$_4$OH, MeOH; (h) NaIO$_4$, MeOH, H$_2$O, −10°; (i) NaBH$_4$, MeOH, H$_2$O, −10°; (j) Me$_2$C(OMe)$_2$, p-TsOH, Me$_2$CO, rt; (k) TBDPS-Cl, DMF, rt; (l) p-TsOH, MeOH, rt; (m) Pb(OAc)$_4$, EtOAc, rt; (n) PDC, DMF, rt; (o) Pb(OAc)$_4$; (p) DCE, N-Ac,O-TMS-cytosine, TMSOTf, then separate isomers.

followed by borohydride and the resulting diol **31** immediately protected as the isopropylidine ketal. The remaining alcohol in **32** was protected as the silyl ether and the ketal group removed by acidic hydrolysis. Another diol cleavage sequence, this time using Pb(OAc)$_4$, was necessary and the intermediate aldehyde oxidised with pyridinium dichromate without S-oxidation. Oxidative decarboxylation of the acid **33** provided the key intermediate **34**, chirally intact at the 2-position, and this was condensed with O-TMS, N-Ac-cytosine which could be taken on to 3-TC as described above. Once again, only limited regio-control was witnessed in the condensation reaction with the desired β-anomer being favoured by a 2:1 ratio.

References:

Zidovudine:
J. P. Horwitz, J. Chua and M. Noel, *J. Org. Chem.*, 1964, **29**, 2076
R. P. Glinski, M. S. Khan, R. L. Kalamas and M. B. Sporn, *J. Org. Chem.*, 1973, **38**, 4299
C. K. Chu et al., *Tet. Lett.*, 1988, **29**, 5349
M. E. Jung and J. M. Gardiner, US Patent 1993, 5,220,003

Lamivudine:
B. Belleau and N. Nguyen-Ba, US Patent, 1991, 5,047,407

D. C. Humber at al, *Tet. Lett.*, 1992, **33**, 4625

J. Saunders and J. Cameron, *Medicin. Res. Rev.*, 1995, **15**, 497

D. Liotta and W. B. Choi, US Patent, 1993, 5,204,466

J. W. Beach et al., *J. Medicin. Chem.*, 1992, **57**, 2217

10 Quinolones as antibacterial DNA gyrase inhibitors

10.1 Introduction

During the preparation of aza-analogues of the anti-malarial agent, chloroquine **1** (Scheme 10.1), the quinolone **2** was isolated and displayed good antibacterial properties, particularly against some Gram negative bacteria. Further structure activity studies resulted in the synthesis of nalidixic acid **3**, the prototypic member of this class. Since nalidixic acid was introduced in 1963, hundreds of other analogues have been made and several have been brought to market. Of the first generation, nalidixic acid and its analogues have found most utility in the treatment of infections of the urinary tract where the drug seems to concentrate. To improve the spectrum of activity and to overcome increased levels of resistance, newer agents were sought and the second generation quinolones exemplified by norfloxacin **4**, ciprofloxacin **5** and ofloxacin **6** are now widely used.

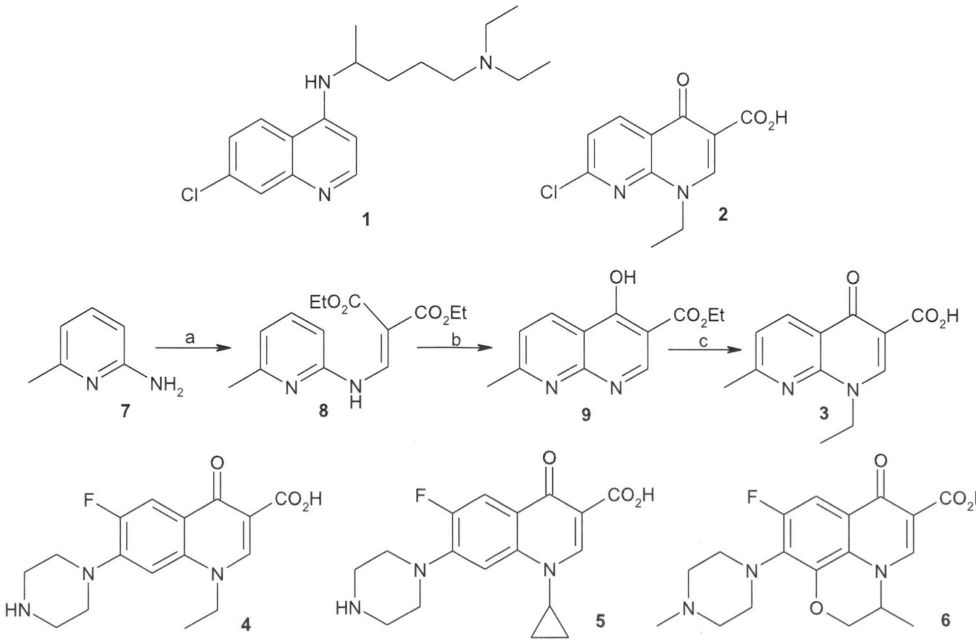

Scheme 10.1. Reagents: (a) 110°, 40 min; (b) Ph$_2$O, reflux, 10 min; (c) KOH then EtI, K$_2$CO$_3$.

Nalidixic acid was readily synthesised starting with a reaction of 2-amino-6-methylpyridine **7** with diethyl ethoxymethylenemalonate. The intermediate enamine **8** was cyclised at high temperature to the hydroxypyridine 8 hydrolysis and N-alkyaltion of which afforded **9**.

All drugs of this class, unlike the cephalosporins and penicillins which inhibit bacterial cell wall biosynthesis, act as potent inhibitors of bacterial DNA synthesis. Bacterial DNA exists as a negative, right-handed supercoil as a result of the ends being covalently closed to form circular DNA and this 'compact' form of DNA is required so that the genome can be folded into the cell. In addition, the enzyme-catalysed make and break reactions (of double stranded DNA) are required during the replication of DNA and its translation into RNA. The processes are mediated by the poly-functional enzyme, DNA-gyrase and involves sequential DNA cutting, nicking and reunion coupled to the hydrolysis of ATP to produce the high energy superhelix. Quinolone antibiotics inhibit DNA gyrase and thus prevent the topological changes needed for DNA replication and RNA transcription and leaves a relaxed form of bacterial DNA which can not yield the correctly folded chromosome.

10.2 Norfloxacin

3-Chloro, 4-fluoroaniline was heated with the Michael acceptor **10** (Scheme 10.2), to give the adduct **11** which was not isolated but immediately cyclised by refluxing in diphenylether. The 4-hydroxypyridine **12** was the only product, the alternative cyclisation presumably being inhibited by the presence of the bulky chloro- group, and this was preferentially alkyated on nitrogen by reaction with EtI in the presence of potassium carbonate. The pyridine ester **13**, now locked in the pyridone form, was hydrolysed to the carboxylic acid and the chlorine displaced by piperazine presumably by an addition – elimination reaction with the molecule behaving as a vinylogous acid chloride.

Scheme 10.2. Reagents: (a) 130°, 2 h; (b) Ph$_2$O, reflux 1 h; (c) EtI, DMF, K$_2$CO$_3$, 90°, 10 h; (d) 2N-NaOH, reflux, 2 h; (e) piperazine, 130°, 5 h.

Scheme 10.3.   Reagents: (a) EtOH, −60° then rt, 16 h; (b) PhMe, 10°, Et₃N, then reflux, 6 h; (c) PhMe, KOBuᵗ, 16 h, rt; (d) DCM, Br₂, 10°, 10 min., then Et₃N, 3 h; (e) KOH, H₂O, 90° 30 min; f) DMSO, 140°, 1.5 h, piperazine.

10.3   Ciprofloxacin

It was necessary to use an alternative process for the synthesis of ciprofloxacin **3** with the N-cyclopropyl substituent being incorporated at an earlier stage (Scheme 10.3). β-Cyclopropylamino propionic acid methyl ester was prepared from methylacrylate and cyclopropylamine and then heated with **14** to give **15** as the major product. This material was cyclised to **16** and oxidised by treatment with bromine followed by elimination to the quinolone **17**. Once again, nucleophilic displacement of the chloro group by piperazine afforded the desired drug **5**.

During the 1990's, the concept of combinatorial chemistry, originally developed for peptides and peptidomimetics, was applied to the synthesis of drug like libraries of compounds. An early demonstration of the power of this new technology was provided by a solid phase approach to quinolones which included ciprofloxacin (Scheme 10.4). The ester **18** was *trans*-esterified with the benzyl alcohol supported on a Wang resin to give the β-keto ester **19** now attached to the resin. Activation of the β-keto ester with dimethylformamide dimethylacetal and capture of the intermediate with any suitable primary amine ('monomers', only 2 were used for this demonstration) gave the corresponding enamides exemplified by **20**. Both of these enamides were cyclised in a reaction catalysed by tetramethylquanidine to afford the quinolone (e.g. **21**) which was then subjected to nucleophilic aromatic substitution. Four different secondary amines were selected; for ciprofloxacin **5** the required amine was piperazine. All resin bound esters were cleaved from the support by exposure to TFA and the products purified by HPLC; overall yields varied from 6–24 %. Given that there are over 1000 primary and secondary amines commercially available, this methodology could, in principle, give access to 1 million analogues. Of course, one would carefully select a small subset of monomer inputs based on computational screening of

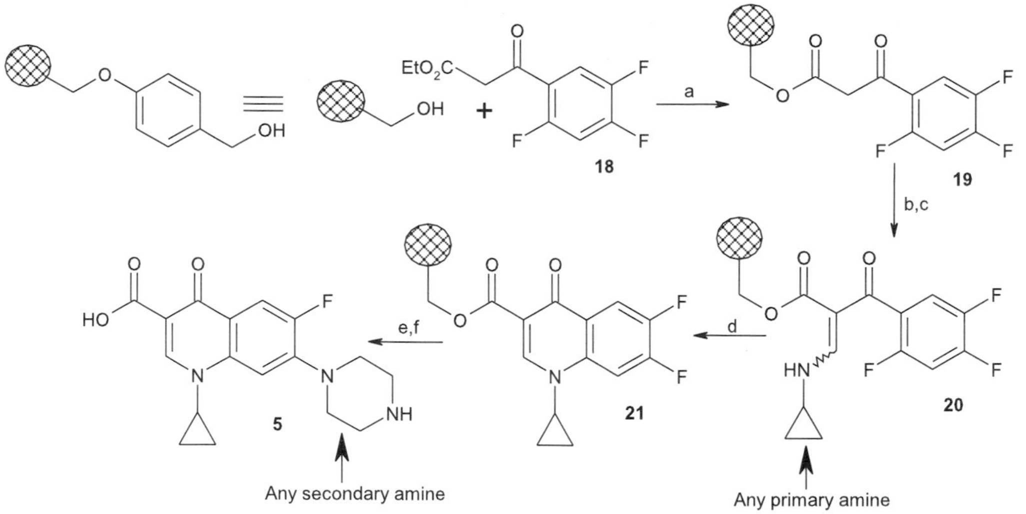

Scheme 10.4. Reagents: (a) DMAP, PhMe, 110°, 18 h; (b) (MeO)$_2$CHNMe$_2$, THF, 25°, 18 h; (c) cyclopropylamine, THF, 72 h, 25°; (d) TMG, DCM, 55°, 18 h; (e) piperazine, N-methylpyrrolidinone, 110°, 4 h; (f) TFA, DCM, 25°, 1 h.

the final virtual products so that a maximally informative library would be available for biological screening.

10.4   Ofloxacin and Levofloxacin

Unlike other quinolone antibiotics, ofloxacin has an additional ring (a 1,4-oxazine) and a stereogenic center which require attention during the synthesis; the drug was developed as the racemate but the more active enantiomer, the (−)-form is now sold as levofloxacin. The first synthetic route started from

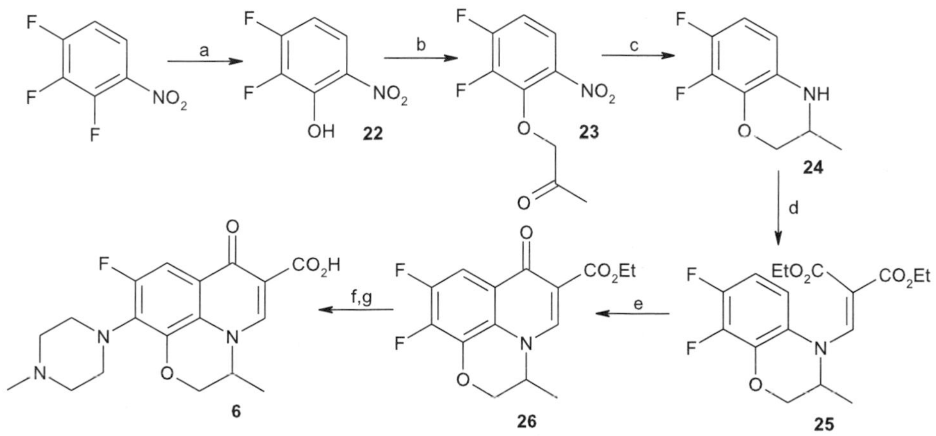

Scheme 10.5. Reagents: (a) DMSO, 10% KOH, 2 h, rt; (b) Me$_2$CO, Cl-CH$_2$COMe, K$_2$CO$_3$, KI, reflux, 4 h; (c) EtOH, Ra-Ni, H$_2$; (d) EtOC = C(CO$_2$Et)$_2$, 140°, 1 h; (e) PPE, 140°, 1 h; (f) conc. HCl, HOAc, reflux, 3 h; (g) DMSO, N-Me-piperidine, 12 h, 140°.

2,3,4-trifluoronitrobenzene which was subjected to alkaline hydrolysis to give **22** as the major product. The outcome of this reaction is predicated by synergistic activation of the 2-F by the ortho-nitro and -fluoro groups but competitive reaction at the 4-position necessitated a chromatographic separation.

O-Alkylation of the phenol with chloroacetone gave **23** which was reduced and cyclised spontaneously to the dihydro benzoxazine **24**. Michael addition of this amine to diethyl ethoxymethylenemalonate under forcing conditions gave the adduct **25** ready for cyclisation to the pyridone using polyphosphoric ethyl ester. Hydrolysis of the ethyl ester **26** under acidic conditions and displacement of 7-fluoro group (quinolone numbering) by N-methyl piperidine afforded **6** without the formation of the alternative regioisomer.

An intramolecular cyclisation involving a nucleophilic attack of the hydroxyl group already attached to the preformed quinolone on to the substituted phenyl ring provided an efficient alternative (Scheme 10.6). The keto ester **27** was readily available from **28** by reaction with ethoxymagnesium malonate and this was reacted with N-methylpiperazine. Both 2-F and 4-F are activated by the electron withdrawing ketone carbonyl but it was found that regioselective displacement of the desired 4-F occurred when the reaction was performed in acetonitrile in the presence of a weak base such as bicarbonate. The resulting amine **29** was treated with dimethylformamide dimethylacetal to give the intermediate **30** which was immediately reacted with 2-amino-1-propanol. The resulting enaminoketone **31** was treated with KF in DMF with the anticipation that a one step conversion to **6** would occur as was the case in the model system when the piperazine ring was

Scheme 10.6.   Reagents: (a) EtOMgCH(CO$_2$Et)$_2$, then hydrolysis; (b) N-methylpiperazine, MeCN, NaHCO$_3$, 3 h, reflux; (c) (MeO)$_2$CHNMe$_2$, PhMe, 110°, 1 h; (d) HOCH$_2$CH(CH$_3$)NH$_2$, EtOH, 0°, then 1 h, rt; (e) dry KF, DMF, 150°, 3 h; (f) NaH, dioxane, 80°, 1 h.

replaced by F. Instead, the intermediate **32** was isolated presumably because the electron donating character of the piperazine ring reduces reactivity towards nucleophilic displacement. However, the synthesis was completed by treatment of **32** to a stronger base, NaH, and indeed, exposure of **31** to NaH gave **6** directly.

Several methods have been reported for the preparation of the enantiomers of ofloxacin. It is now know from X-ray analysis of the separated enantiomers that the more active isomer is the (S)-(−)-form and this is marketed as levofloxacin. Perhaps the route most ammenable to large scale production involves an enzyme catalysed hydrolysis of **34** selective for one isomer over the other (Scheme 10.7). Starting from 2,3-difluoro, 6-nitrophenol, 33 was available using chloromethyl acetoxymethylketone as the alkylating agent and the product reduced with Raney Nickel. The resulting amine cyclised *in situ* to give the benzoxazine **34** as a mixture of enantiomers. This mixture was incubated with lipoprotein lipase in a biphasic reaction medium, an organic phase of hexane and benzene and an aqueous phase maintained at physiological conditions. For larger scale preparations, the enzyme was adsorbed on to an Amberlite resin and re-used over several cycles. It was discovered that the (R-(+)-isomer was hydrolysed to the primary alcohol **35** leaving the (S)-isomer **36** unchanged and these products were easily separated by column chromatography. The enantiomerically pure acetate was hydrolysed to **37** and then treated successively with thionyl chloride and

Scheme 10.7. Reagents: (a) ClCH$_2$COCH$_2$OAc, Me$_2$CO, K$_2$CO$_3$, rt, 30 min; then KI, reflux 4 h; (b) MeOH, Ra-Ni; (c) PhH, hexane, phosphate buffer (pH 7.0), lipoprotein lipase, 37°, 6 h then column chromatography; (d) 1N-KOH, THF, 30 min, rt; (e) pyridine, SOCl$_2$, 0°, then 50°, 40 min; (f) NaBH$_4$, DMSO, 85°, 1 h.

Scheme 10.8.   Reagents: (a) MeCN, 0–5°, 8 h then rt 1 h; (b) Et$_3$N, DMAP, DCM, MeCOCl, 0°, 30 min.; (c) Et$_3$N, DCM, MeCN, 0°, 30 min; (d) THF, KOH, 0–25°, 1 h then add H$_2$O, 2 h, reflux.

borohydride to effect conversion to the methyl group seen in the final drug product. The methyl-substituted benzoxazine **38** was taken through to **(S)-(–)-6** using the chemistry described in Scheme 10.5.

Since both (S)- and (R)-alaninol are readily accessible from the alanine, methods were investigated which used this synthon. Such an approach started with a Michael reaction between (S)-(+)-2-aminopropanol and ethyl propiolate (Scheme 10.8) to give after protection of the primary alcohol as the acetate, the acrylate **39**. This was then reacted with tetrafluorobenzoyl chloride in the presence of base to yield the adduct **40**. As indicated above, it was possible to effect a double cyclisation of **40** by first unmasking the primary alcohol under mildly basic conditions followed by more forcing conditions give **41** without isolation of intermediates. It was subsequently shown that the same homochiral starting material [either (S)- or (R)-alaninol] could be incorporated into (S)- or (R)-**6** more directly by using the β-keto ester **27** as the starting material followed by reaction with diethyl ethoxymethylene malonate and then following the chemistry outlined in Scheme 10.6.

References

Nalidixic Acid
G. Y. Lesher et al., *J. Medicin. Chem.*, 1962, **5**, 1063

Norfloxacin
T. Irikura, US Patent, 1981, 4,292,319
H. Koga et al., *J. Medicin. Chem.*, 1980, **23**, 1358

Ciprofloxacin
K. Grohe, H. J. Zeiler and K. G. Metzger, US Patent, 1984, 4.670.444
A. A. MacDonald et al., *Tet. Lett.*, 1996, **37**, 4815

Ofloxacin
US Patent, 1982, 4,382,892

I. Hayakawa, T. Hiramitsu and Y. Tanaka, *Chem. Pharm. Bull.*, 1984, **32**, 4907

H. Egawa, T. Miyamoto and J-I. Matsumoto, *Chem. Pharm. Bull.*, 1986, **34**, 4098

Levofloxacin
I. Hayakawa et al., US Patent 1991, 5,053,407

Y. Kim, S. B. Kang and S. Park, US Patent 1996, 5,539,110

Abbreviations

AIBIN	Azobisisobutyronitrile
BINAP	2,2′-bis (diphenylphosphino) -1, 1′-binaphthyl
CAN	ceric ammonium nitrate
CDI	carbonly diimidazole
DBA	dibenzanthracene
DCCI, DCC	dicyclohexylcarbodiimide
DCE	1, 2-dichloroethane, ethylene dichloride
DCHA	dicyclohexylamine
DCM	dichloromethane, methylene dichloride
DIBAL	diisobutylaluminium hydride
DMA	N, N-dimethylacetamide
DMAP	dimethylaminopyridine
DME	dimethoxyethane
DMF	dimethylformamide
DMS	dimethylsulphide
DMSO	dimethylsulphoxide
DPPA	diphenylphosphoryl azide
h	hour
HBT	1-hydroxybenzotriazole
IPA	iso-propanol
LAH	lithium aluminium chloride
Mcpba	m-chloroperbenzoic acid
MsCl	methanesulphonyl chloride
NBS	N-bromosuccinimide
Ni(COD)2	bis(1, 5-cyclooctadiene) nickel(0)
NMP	1-methyl-2-pyrrolidinone
Psi	pounds per square inch
Py	pyridine
Ra-Ni	Raney Nickel
Red-Al	sodium bis(2-methoxyethoxy) aluminium hydride
rt	room temperature
TBAF	t-butylammonium fluoride
TBDPSCI	t-butyl-diphenylsilyl chloride
TFA	trifluorooacetic acid
THF	tethrahydrofuran
TMEDA	N, N, N′, N′, -tetramethylethylenediamine
TMG	tetramethylguanidine
TMSI	trimethylsilyl iodide
TMS-triflate (-OTf)	trimethylsiyl trifluoromethanesulphonate

Index

agonist 33, 55
AIDS 73
allergy 63
alprazolam 55, 59, 60
amlodipine 24
angiotensin converting enzyme 1, 4, 18
angiotensin II 4, 14
antagonist 34, 55
antibacterial 82
anti-viral agents 73
anxiolytic 1, 55, 56
Arbuzov 15
aspirin 63
astemizole 63, 71

benazepril 5
benzodiazepine 1, 54, 55, 60
blood brain barrier 63
Buchwald and Hartwig 71

calcium channel 23, 24, 29
calmodulin 23
captopril 4, 5, 6, 7, 10, 14
carboxypeptidase A 5
cetirizine 63, 68, 70
chlordiazepoxide 54, 56
chloroquine 82
cimetidine 34, 35
ciprofloxacin 82, 84
clomipramine 45
cytidine deaminase 78

Darzen 28
depression 1, 45
diazepam 55, 56, 57, 60
Diels-Alder 49
diltiazam 23

DNA-gyrase 83

enalapril 5, 8, 18
enantioselective 11, 28, 49, 70, 76, 79
enzyme 2, 5, 18, 39, 40, 73, 78, 83, 87

famotidine 34, 37
fexofenadine 63, 64, 66
Finklestein 69
flumezanil 55, 60
fluoxetine 45, 47, 48
fosinopril 5, 14
Friedel-Crafts 48, 49, 67

GABA$_A$ 55
GPCR 33, 63
Grignard 57, 64, 66, 67, 68

H$^+$/K$^+$ ATPase 39, 40
Hantzsch 23, 24, 25
histamine 1, 33, 34, 39, 63
HIV 1, 18, 73
homochiral 25, 29, 79, 88
homoconjugate addition 50
hypertension 1, 3, 23, 24
hypnotic 55, 60

imipramine 45
intramolecular radical cyclisation 12

Knoevenagel 24

lamivudine 74
lansoprazole 43

lisonopril 5
loratidine 63, 67, 68
lorazepam 55, 56
losartan 18, 19, 21

Mannich 35, 36, 45, 61
Meldrum's acid 25
mepyramine 33
metabolite 18, 21, 55, 63, 67
Michael 12, 36, 51, 75, 83, 86, 88
Mitsunobu 60

nalidixic acid 82
nifedipine 23, 24
noradrenaline 23, 45, 47
norfloxacin 82
nucleoside 1, 73

ofloxacin 82, 85, 87
omeprazole 39, 40, 41, 42, 43
omeprazole cycle 40

paroxetine 50, 53
penicillin 63
pharmacokinetic 22
Pictet Spengler 9
Polonovsky 56
prodrug 4, 5, 14, 15, 21
protease 18
proton pump 39

quinapril 5, 8
quinolone 83

ramipril 5, 11, 12

ranitidine 34, 35, 36
receptor 1, 5, 23, 33, 34, 55, 60
resolution 6, 11, 12, 29, 46, 68
reverse transcriptase 1, 73

Ritter 68

Schmidt ring enlargement 10
seasonal rhinitis 63
serotonin 1, 45, 46, 50
sertraline 48, 49, 50
Sharpless 27, 53, 76
stereogenic 11, 12, 15, 23, 31, 42, 66, 85
Stille 49
Stobbe 48
synthon 35, 88

temazepam 55, 56
terfenadine 63, 64

ulcers 33, 39
Ullmann biaryl synthesis 19, 20

valsartan 19, 21
verapamil 23, 29, 30, 31
Von Braun 45
Vorbruggen 75, 76, 78

Wadsworth-Emmons 67
Wang resin 84

zidovudine 73
zolpidem 55, 62